Software-Hardware Integration in Automotive Product Development

Other SAE books of interest:

Multiplexed Networks for Embedded Systems
By Dominique Paret
(Product Code: R-385)

Automotive Software Engineering
By Joerg Schaeuffele and Thomas Zurawka
(Product Code: R-361)

Vehicle Multiplex Communication
By Christopher A. Lupini
(Product Code: R-340)

For more information or to order a book, contact:
SAE International
400 Commonwealth Drive
Warrendale, PA 15096-0001 USA

Phone: +1.877.606.7323 (U.S. and Canada only)
or +1.724.776.4970 (outside U.S. and Canada)
Fax: +1.724-.776.0790;

Email: CustomerService@sae.org;
Website: books.sae.org

Software-Hardware Integration in Automotive Product Development

Edited by John Blyler

INTERNATIONAL

Warrendale, Pennsylvania, USA

400 Commonwealth Drive
Warrendale, PA 15096-0001 USA

E-mail: CustomerService@sae.org
Phone: 877-606-7323 (inside USA and Canada)
        724-776-4970 (outside USA)
Fax: 724-776-0790

ISBN 978-0-7680-8052-0
Library of Congress Catalog Number 2013949773
SAE Order Number PT-161
DOI 10.4271/ PT-161

Information contained in this work has been obtained by SAE International from sources believed to be reliable. However, neither SAE International nor its authors guarantee the accuracy or completeness of any information published herein and neither SAE International nor its authors shall be responsible for any errors, omissions, or damages arising out of use of this information. This work is published with the understanding that SAE International and its authors are supplying information, but are not attempting to render engineering or other professional services. If such services are required, the assistance of an appropriate professional should be sought.

To purchase bulk quantities, please contact
SAE Customer Service
e-mail: CustomerService@sae.org
phone: +1.877.606.7323 (inside USA and Canada)
+1.724.776.4970 (outside USA)
fax: +1.724.776.0790

Visit the SAE International Bookstore at
books.sae.org

Table of Contents

1.0 Introduction

All of the papers referenced in this compendium attest to the difficulties of the integration, verification, and validation (IVV) phase of the automotive hardware and software system life cycle. Yet, the future of automotive product development lies in the timely integration of these chiefly electronic and mechanical systems. How can the automotive industry overcome these difficulties?

These carefully selected papers demonstrate how leading companies, universities, and organizations have developed methodologies, tools, and technologies to integrate, verify, and validate hardware and software systems. The integration activities cross both product type and engineering discipline boundaries to include chip-, embedded board-, and network/vehicle-level systems. Integration, verification, and validation of each of these three domains will be examined separately and then together.

Each of the domains uses identical terms to describe key concepts, elements, and processes (e.g., modeling, hardware-software, and systems). Before proceeding, these terms and others must be defined to avoid confusion, and set the proper context for the rest of the book.

IVV is that portion of the complete product or system life-cycle phase captured in the right-hand side (RHS) of the V-diagram model (see Fig. 1). This RHS portion is sometimes called the "build, test, and deliver phase [1]." The model's left-hand side is where the architectural design is functionally partitioned and decomposed into manageable subsystems and components. At the lowest levels of decomposition—i.e., the bottom of the "V"—the resulting subsystems are ready to be recomposed into an integrated whole. This integration phase is where verification, validation, and test of all the hardware and software occur.

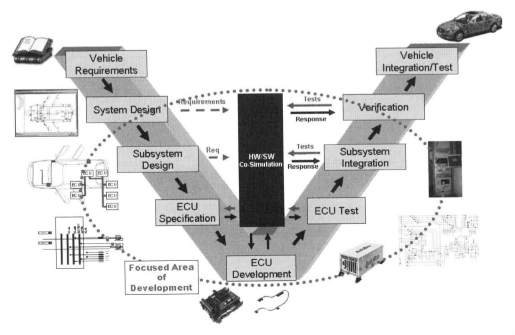

Fig. 1 Model-based V-diagram showing areas of focus (SAE Paper 2008-21-0043).

1

What is meant by verification and validation? How does testing fit into all these activities? Verification is the process of ensuring that the system or product was built correctly, as specified by the design. Similarly, validation ensures that the correct system or product was built (i.e., that the correct problem was addressed). Verification and validation are accomplished by testing at every level, from individual subsystems/units through the integration of the complete system.

The current state of the art is to integrate, verify, validate, and test automotive hardware and software with a complement of physical hardware and virtual software prototyping tools. The growth of sophisticated software tools, sometimes combined with hardware-in-the-loop devices, has allowed the automotive industry to meet shrinking time-to-market, decreasing costs, and increasing safety demands. It is also why most of the papers in this book focus on virtual systems, prototypes, and models to emulate and simulate both hardware and software. Further, such tools and techniques are the way that hardware and software systems can be "co-verified" and tested in a concurrent fashion.

Are physical hardware and software prototyping tools the best way to perform IVV and test? Probably not, according to Bill Chown, product marketing director for the system-level engineering division at Mentor Graphics. "It is very hard to do worse case or statistical analysis with a specific instance or even limited set of instances of a piece of hardware," observed Chown. What can be done?

A model-driven design, implementation, and test approach has gained growing acceptance over the years [2, 3]. A model-driven development (MDD) process supports verification of the model at each stage of design and thus reduces reliance on late-stage physical integration and testing as the primary means to validate the system, notes Chown.

Another benefit of model-driven development is its supports of the design and verification of mechatronic automotive systems and the growing need for cross-disciplinary collaboration. Mechatronics is a development process that requires a combination of mechanical, electrical/electronic, control, and computer engineering disciplines.

"Vehicle electronics and embedded systems currently represent 20 to 40% of global development costs and are projected to grow at 9% a year," noted Michael Lalande, Director of Transportation & Mobility, Dassault Systèmes. "To control development costs and remain competitive requires the integration of mechanical, electronic, and software components into one virtual and collaborative environment." But a virtual environment requires extensive modeling of all hardware and software components and processes. However, once created, these virtual models can be used again and again.

The reuse of mechanical and electronic hardware as well as software code, libraries, and models is a critical element of today's design and integration activities. Reuse of such intellectual property (IP) is a prerequisite to meet performance, time, and cost demands of modern hardware-software systems.

The goal of this compilation of expert articles is to reveal the similarities and differences between the integration, verification, and validation (IVV) of hardware and software at the chip, board, and network levels. This comparative study will reveal the common IVV thread among the different, but ultimately related, implementations of hardware and software systems. In so doing, it supports the larger systems engineering approach for the vertically integrated automobile—namely, that of model-driven development.

2.0 Embedded Board-Level Hardware-Software

In the past, the greatest obstacle in hardware-software development was one of time-dependence. Software could not be started until the hardware platform on which the software ran was finished. Today, faster and less expensive computational platforms and model-based methodologies ensure that software can be designed and verified in sync with the hardware (i.e., co-designed and co-verified).

Such co-development approaches require the use of simulation models and virtual prototypes. Software designers have long appreciated the benefits of virtual prototypes, which ran at increasing speeds and complexity thanks to the increasing computational benefits of Moore's law. These same performance benefits are now enjoyed by hardware designers and integrators. (See Paper 2008-21-0043, "Adaptation of a 'Virtual Prototype' for Systems and Verification Engineering Development.")

Standards have also helped the verification and validation (V&V) effort. First, they provide a set of specifications for moving between levels of abstraction (e.g., untimed architectural design through detailed, timing-specific hardware models). Secondly, at a process level, standards provide a common way for V&V to be handled. An example of the latter is ISO 26262, an international functional safety standard. It is an adaptation of the IEC 61508 specification for electrical and electronic systems in the road vehicle industry. While focused on the development of in-vehicle application software development, work is being done to implement this approach with model-based developments. (See Paper 2009-01-2929, "Verification and Validation According to IEC 61508: A Workflow to Facilitate the Development of High-Integrity Applications.")

Interestingly, the success of ISO 26262 might serve as a working model for V&V in other industries such as chip design and electronic design automation (EDA) tools.

The implementation of hardware-software systems is significantly different in each of the three major domains or abstraction-levels (i.e., chip-, embedded board-, and the network/vehicle-level). For example, system-on-chip (SoC) developers understand hardware in terms of transistor-based devices, and verification is increasingly handled as part of a bundled IP reuse package.

These IP packages are typically the hardware cores and software stacks needed for popular interface input/output (I/O) subsystems such as standard PCIe bus or Ethernet interface. Additionally, such IP subsystems include both the hardware and software and the verification and integration suites need to ensure that the interface (e.g., Ethernet) is properly integrated and tested within the larger system, such as the rest of the automobile electronics [4].

At the next level of abstraction beyond the chip are embedded board-level developers who view hardware in terms of microprocessor/microcontroller components, specialized chips, memory, and interface subsystems. Software is seen in terms of firmware or device drivers, and a hardware-centric real-time operating system (RTOS) that may be part of a high-level operating system (OS).

Finally, boards are connected into larger network-based systems that include the entire vehicle. In automotive terms, multiple embedded control units (ECUs)—consisting of the previously mentioned components—are distributed across one or more of the SAE International automotive network classifications such as Local Interconnect Network (LIN), Controller Area Network (CAN), Media Oriented Systems Transport (MOST), FlexRay, and Ethernet [5]. Here, board-level embedded hardware systems become a subsystem to modules and larger network systems. Board-level firmware and operating systems (OSs) become secondary to network operating systems (NOS), middleware, and finally end-user applications.

Naturally, there could be an even higher level above networks that connect the automobile to other transportation systems. But in the context of a single automobile, hardware and software integration occurs at these three levels. Most automotive experts agree that a vertically integrated system for designing and verifying all hardware and software subsystems is vital for today's safety- critical, consumer- driven automotive market.

3.0 Chip-Level Hardware-Software

The essence of hardware and software co-design/co-verification is parallelization (i.e., enabling both design and verification to proceed at roughly the same time early in the design process and continuing through verification, testing, and final product integration). Achieving this goal requires the use of virtual systems to investigate trade-off options between hardware—such as microprocessors, memory, and interfaces—and software implementations before the hardware is fully designed. These trade-off studies performed in the design phase provide the basis for verification specification and debugging/testing during the integration phase [6].

What is a virtual system? It is a method and set of tools used for the architectural evaluation of hardware and software trade-offs prior to the creation of actual hardware or "hardware realization." In Paper 2006-01-0170, "Hardware/Software Design and Development Process," a virtual automotive airbag system is created to explore the interactions between a 32-bit microcontroller coupled with seven application-specific integrated circuit (ASIC) devices and the associated software.

This example highlights the main challenge in exploring hardware and software systems: namely, that electronic hardware models must operate in a timed, fully cycle-accurate fashion to properly represent the hardware. One the other hand, software requires fast high-level models that do not need to be cycle-accurate. The virtual airbag example was a combination of timed functional, cycle-accurate, and cycle-approximate simulation.

The EDA tools industry has been promoting a high-level abstraction modeling approach called electronic system level design (ESL). One standards-based approach to ESL is transaction-level modeling (TLM), which makes it easier for chip designers to build virtual platforms for hardware/software co-design and co-verification. TLM is a modeling technique that integrates both cycle-accurate hardware models and untimed software models.

The co-design of hardware and software provides the basis for the roughly simultaneous co-verification effort, matched to the level of abstraction of each phase of the V-diagram life cycle. In other words, high-level architecture is matched with high-level verification, while unit-level design results in unit-level verification and test procedures. For example, software simulators and the virtual processor models (VPMs) used in design will integrate with debuggers appropriate to a given level of abstraction (e.g., the system-on-chip circuits).

Today, most models are used to simulate digital computing and communication microprocessor-based systems. But a new class of devices is emerging that relies upon analog, input-intensive microcontrollers; mixed-signal-conditioning circuits; and the related software prevalent in mobile, sensor-rich devices.

The explosive growth of these mobile communication devices, often associated with the Internet-of-Things (IoT), has spawned the creation of smart sensors. Sometimes referred to as sensor fusion, these smart devices combine analog sensor hardware with microcontroller-based software. Tiny mixed-signal devices interface with a microcontroller to control some aspect of the physical system such as motion, speed, temperature, etc.

Smart fusion technology supports the growing demands for information-based infotainment and automotive control applications. However, as noted in Paper 2004-01-0718, "Using VHDL-AMS as a Unifying Technology for HW/SW Co-verification of Embedded Mechatronic Systems," this added dimension of software presents a significant challenge to the design, integration, and verification of the overall system.

To address these challenges, a mixed-signal hardware modeling language was used. The VHSIC Hardware Description Language (VHDL)—Analog Mixed Signal (AMS) extension provides a behavioral model for physical sensor and wireless systems. This is yet another example of a system prototype, one that uses a model of the microcontroller to execute the software instructions that control the overall operation and performance. The trick here is to incorporate the software algorithm into the flow of the simulation.

It can be difficult to debug and verify the operation of the software within the context of the complete system (e.g., in an embedded, smart-sensor mechatronic system such as a Tire Pressure Monitoring System—TPMS). Often, the actual system performance is not known until the first physical prototype is built. But by that time in the design process, correcting unanticipated hardware and software interactions or software bugs is often very expensive.

One solution is to use a hardware modeling language—such as VHDL—for successive verification of the design, from design specification through verification and final certification. The referenced paper (2004-01-0718) provides a dc motor servo loop example to illustrate how to apply successive verification techniques using VHDL-AMS.

We've seen how virtual system prototypes are important for both the co-design and co-verification of hardware and software systems. Many automotive subsystems (e.g., powertrain and safety) contain an ever-increasing amount of processing cores. Understanding, analyzing, and verifying the behavior of all processing platforms requires a virtual prototyping approach to ensure the safe operation of software within the system.

Cost-effective chip-level architectural exploration, and later verification, rely upon IP cores. Verification IP blocks have become more critical thanks to the increase in subsystem interface I/O and memory IP blocks—as was discussed earlier [7].

The use of accurate virtual hardware-software prototype models depends upon robust modeling methodologies and languages, e.g., C++ for untimed, architectural software modeling, and VHDL or Verilog for low-level, timed hardware modeling. (See Paper 2005-01-1342, "Virtual Prototypes as Part of the Design Flow.") Trade-off analyses are then performed to determine the optimal implementation of functionality on either the hardware and software, based on key technical performance measures (TPMs), such as speed, system complexity, and the accuracy of results.

Architectural exploration and associated verification requires fast and accurate models. The models are tailored to the implementation type (e.g., untimed for software and timed for detailed hardware modeling). Simulation tools must be capable of drawing from a broad model library, and reuse of existing components and IP.

4.0 Network/Vehicle-Level Hardware-Software

So far, we have demonstrated techniques for IVV and testing of hardware-software systems at the chip and board levels. But modern automobiles are a collection of networks that connect components and boards throughout the vehicle. How is V&V and testing accomplished at the vehicle level of networked systems?

Paper 2008-21-0041, "To Test the Need and the Need to Test—Testing the Smart Controller Network for the Chassis of Tomorrow," examines the benefits of using hardware-in-the-loop (HIL) simulation, which uses emulators to convince the embedded system under test that it is operating with real-world inputs and outputs. Typically, automotive suppliers test their own modules and components, such as an ECU. Then, the original equipment manufacturers (OEMs) test the ECU as part of the larger vehicle network. The paper describes the main differences between component and network HIL tests for validation of automotive electronics. Additionally, it highlights improvements in the network test bench approach to validate and verify connected subsystems such as the automatic transmission, torque on demand transfer case, torque vectoring axle drive, electric power steering, leveling control, as well as a brake control system (ESP). Key innovations of this network test bench are also covered.

Until recently, the actual physical vehicle was where components and boards were first integrated and verified. This mirrored the hardware-software approach already discussed (e.g., the delay of software development until the actual electronic hardware was built). Today, both suppliers and OEMs perform HIL tests, but from quite different points of view. For example, suppliers test the ECU as a component, while OEMs test the ECU from a network perspective. On the right-hand side of the V-diagram life cycle, suppliers test on the bottom half, while OEMs test and further recompose the system on the upper half, all the way up to the finished product.

The reuse of test suites and verification IP enable efficiently meeting shrinking time-to-market (TTM) deadlines and costs. Unfortunately, the reuse of complete test cases will only typically work within the same product line (supplier's view) and/or vehicle platform (OEM's view). Distinguishing between these two positions is a critical part of the reuse strategy.

Regardless, the system design and test specifications must be linked to each other and kept up-to-date throughout the entire development process. This linkage between design and verification at all abstraction levels is the reason for the popularity of the V-diagram life-cycle model discussed earlier.

Many automotive body control applications are handled by numerous ECUs that are spread throughout one or more vehicle networks. Design verification and product validation must occur at the vehicle level for these systems. This requires the availability of both the networks and ECUs during user operation.

The challenge is that ECUs are developed by different suppliers, while the vehicle networks' infrastructure and communication protocols are normally developed by the OEM. The individual ECUs cannot be fully verified and tested until all the ECUs are connected to all the vehicle networks. This event typically happens during the final stages of vehicle-level integration testing, when the cost of fixing errors is at its highest.

One way to solve this delay is a fault-insertion testing strategy, as covered in Paper 2010-01-2328, "A Systems Engineering Approach to Verification of Distributed Body Control Applications Development." A methodology is described for developing a body of control applications based on the concept of executable specification, plant modeling, test case generation using various means, and migration of test cases in the virtual test harness model to ECU-in-the-loop testing environment. Once again, the importance of a virtual systems environment and hardware-in-the-loop simulation testing is emphasized.

Handling hardware and software verification and valuation at the vehicle level requires a model-based methodology. In addition, this approach relies on test-case reuse and virtual testing models—all part of a model-based approach. Test-case models are an effective way to identify and replicate software errors and bugs in the pre-production stage. The methodology is also very useful for a large- scale modeling project in which multiple large models need to go into one microcontroller.

While not a replacement for traditional integrated in-the-loop testing methods, this approach provides an alternative that uses fault-insertion testing, automatic testing, and reusing test cases created in a virtual model-based testing environment.

5.0 Integration and Interface Management

Up to now, the referenced papers have mostly addressed the hardware and software verification, validation, and testing challenges in automotive electronics—from a chip, board (ECU), and vehicle-level network system. But what about the integration effort (i.e., the decomposition activity of the right-hand side of the V-diagram life-cycle process)? How is that best accomplished?

For example, the current automotive entertainment/infotainment environment includes many hardware and software subsystems that are segmented into various and similar technology areas (e.g., radio space, the rear-seat entertainment space, the infotainment space, and the satellite radio space). All of these subsystems are connected through the vehicle-level entertainment network of the car. But this traditional approach is redundant, expensive, not scalable, nor tailored for reuse. Paper 2004-21-0071, "Highly Scalable and Cost Effective Hardware/Software Architecture for Car Entertainment and/or Infotainment Systems," suggests moving away from hardware-intensive, analog interfaces toward an integrated software architecture based on object-oriented methodologies. The hoped-for result is a radio design that offers greater reliability, increased flexibility, faster time-to-market, and lower overall costs through standardization within a scalable hardware and software architecture that spans many market product levels.

Integration—or consolidation—of the various radio spaces into one space means that the overall radio system becomes the responsibility of the major Tier-1 suppliers. From a general system perspective, this will lead to an integrated systems approach where all the media sources, radio frequency (RF) sections, and audio/speech processing are integrated into the radio space. Only the antenna system, video monitor, and remote control are external to the radio head unit. Added features and entertainment media may be offered through common consumer electronics ports, such as Bluetooth, 802.11, USB, SD memory, 1394 ports, and a front-panel stereo audio jack.

From a hardware perspective, the key components such as the processors (including any digital signal processors (DSPs), internal buses, plug and play blocks, and connection portals must be developed as a system.

Conversely, from a software viewpoint, the key components are drivers and programs. For this code to be reusable on other projects and hardware, the functional blocks of the code must be tailored to specific hardware. In this way, the pre-build and pre-tested code is componentized for a particular hardware component (e.g., the software for controlling a particular CD mechanism for a particular family of automobiles). This objective of hardware and software reuse is also the central goal of the AUTomotive Open System ARchitecture, or AUTOSAR—an open and standardized automotive software architecture.

Integration of the automotive RF space is based on a basic hardware-software architectural approach: namely, that common functionality should be partitioned together, connections should be digital, and all things should be scaled accordingly. Additionally, standardization within such a scalable hardware and software architecture would span many product levels.

During the integration phase of the product life cycle, system integrators use interfaces as a primary way to monitor the integration and testing effort. Automotive systems are linked with an ever-increasing amount of vehicle hardware and software interfaces. Ford Motor Company, among other manufacturers, uses a systematic approach to analyze all interfaces for successful system integration. Paper 2008-01-0279, "Analysis of Interfaces and Interface Management of Automobile Systems," presents a step-by-step method for such integration with an idealized example for guidance.

The incorporation of active safety devices such as airbags, ABS, and ESP as standard equipment in all cars has been made possible through the use of electronic controllers and intelligent data networks such as the CAN-Bus. Design and verification of these devices required interface integration among many different engineers and departments. The lack of a robust cross-disciplinary and cross-departmental working environment has been the cause of many new failure modes. These failures could have been lessened if critical interface information was exchanged early in the automotive electronics development process.

The referenced paper (2008-01-0279) presents a useful guide to interface management that includes both the mechanical and electrical components in a vehicle. A variety of tools are used to control the interfaces including boundary diagrams, an interface matrix, and an interface description sheet.

These tools and others are essential to manage the systems integration effort, especially in cross-departmental and cross-discipline environments. The precise control, calibration, and adjustment of numerous vehicle interfaces are a necessity in today's hardware- and software-intensive automotive systems. Detailed interface specifications should be used in technical meetings on a regular basis for design and verification engineers. Systems engineers and architects will use these interface specifics to monitor key technical performance measures such as power, performance, weight, and size [8].

We have already introduced the use of HIL technology as a tool for verification of vehicle-level network systems. In the enclosed paper (2011-01-0443, "Advancements in Hardware-in-the-Loop Technology in Support of Complex Integration Testing of Embedded System Software"), software integration testing is accomplished using HIL hardware consisting of the latest quad-core processors, field programmable gate array (FPGA) based I/O technology, and communication bus systems such as FlexRay.

Also presented are improvements in various software elements such as advanced user interfaces, global positioning system (GPS) information integration, real-time testing, and simulation models.

As mentioned previously, HIL technology has become an integral part of the electronics development process for both suppliers and OEMs to test single and networked ECUs, respectively. Integration testing has to cover a variety of networked ECUs. These variations exist due to platform-and country-specific versions of a vehicle or selectable vehicle options by customers. Likewise, integration testing needs to cover the complexity of vehicle networks consisting of up to 70 different ECUs. Thus, HIL integration testing requires covering a large number of ECUs networked with a number of different buses and protocols covering many interface I/O signals.

Hardware for HIL integration testing must include significant processing power and signal generation and measurement within short cycle times. Modern multi-core processors for new HIL systems require modular inter-core and inter-processor communication. Paper 2011-01-0443, "Advancements in Hardware-in-the-Loop Technology in Support of Complex Integration Testing of Embedded System Software," explains that I/O boards can provide some of the fastest data transfers between the on-board CPU and FPGA. This approach enables very short processing times locally on the I/O board instead of delays resulting from transmission times via an I/O bus to a real-time processor.

Software for HIL testing needs to support functional (behavior) modeling and development of dynamic plant models. Additionally, HIL software must provide real-time testing capabilities that can be used in conjunction with virtual and automation test tools. The real-time script libraries can be used with test automation tools or directly through real-time script languages.

6.0 Summary

The current practice of integrating, verifying, validating, and testing automotive hardware and software relies upon a complement of physical hardware and virtual software prototyping tools.

While this approach is common across chip, embedded board, and vehicle network systems, it is costly and time consuming to implement. Growing hardware and software complexity, the need for cross-disciplinary development teams, and an increase of the global, multi-vendor supply chain adds to the challenge.

A virtual system-based MDD approach addresses many of these challenges. Further, the use of virtual systems and models enables the use of co-verification techniques, which lessen the chance for errors late in the development process. Finally, virtual systems can be used to integrate, verify, validate, and test all levels of hardware and software subsystems, from the chip and embedded board through the complete vehicle network system.

7.0 References

1. Blyler, John, and Gary Ray. *What's Size Got to do with It – Understanding Computer Rightsizing*. IEEE-Wiley Press, 1998.

2. Chown, Bill, and Michelle Lange. "Modernizing System Development: Requirements-Based, Model-Driven Design, Implementation and Test." Mentor Graphics Corp., ERTS, Feb. 2012.

3. Blyler, John. "Model-Driven Development Is Key to Low Power." Chip Design Magazine – JB's Circuit, Dec. 6, 2012. http://www.chipdesignmag.com/blyler/2012/12/06/model-driven-development-key-to-low-power/.

4. Blyler, John. "IP's Silent Presence in Automotive Market." Chipestimate.com, IP Insider, June 3, 2011. http://www.chipestimate.com/blogs/IPInsider/?p=92.

5. Blyler, John. "IP Smoke Testing, PSI5 Sensors, and Security Tagging." Chipestimate.com, IP Insider, April 12, 2013. http://www.chipestimate.com/blogs/IPInsider/?p=1513.

6. Ganesan, Subramaniam, ed. *Automotive Systems Engineering – Approach and Verification (PT-145/4)*. SAE International, 400 Commonwealth Drive, Warrendale, PA, 2011.

7. Blyler, John. "Surprises Abound As Subsystem IP Gains Prominence." Chipestimate.com, IP Insider, February 28, 2013. http://chipdesignmag.com/sld/blog/2013/02/28/surprises-abound-as-subsystem-ip-gains-prominence/.

8. Blyler, John. "Interface management." Instrumentation & Measurement Magazine, Vol.7, Issue 1. IEEE. http://ieeexplore.ieee.org/xpl/login.jsp?tp=&arnumber=1288741&url=http%3A%2F%2Fieeexplore.ieee.org%2Fxpls%2Fabs_all.jsp%3Farnumber%3D1288741.

2008-21-0043

Adaptation of a "Virtual Prototype" for Systems and Verification Engineering Development

M. S. Chandrashekar, B. C. Manjunath, Everett R. Lumpkin and Frank J. Winters
Delphi Electronics and Safety

ABSTRACT

The use of "Virtual Prototyping" technology for software development is becoming increasingly prevalent and critical to Electronic Control Unit (ECU) development in the semiconductor, telecommunications and automotive industries as an effective countermeasure to the increased size and complexity of software code, as well as, the increased quality expectations of the "end-customer". The global expansion of the automotive market is also continuing to drive the need for engineering teams comprised of members from various countries and regions around the globe.

To date, the virtual prototyping simulation environment has been oriented in nature to the software engineer as the primary user of the technology. With the continued advancements in simulation and analysis technology, the capabilities now exist to expand the use of "Virtual Prototyping" to other engineering disciplines in the product development cycle, specifically targeting the systems and Independent Test and Verification (IT&V) engineers.

INTRODUCTION

The trends of increased functionality, improved performance, reduced size and increased complexity continue to evolve in the automotive electronics market. New system architectures are providing the performance and memory capability necessary to keep up with the hardware performance and software growth required by the automotive market trends. All of this technology growth implies a higher product cost and increased engineering effort required to develop these new products.

In order to meet all of the technical and business objectives, corporations must manage project development cost by increasing the productivity and efficiency of engineering activities, reducing developmental spend rates, shortening product development times, and ensuring that the end objectives for product performance and delivery are met to schedule. Making the economic problem more complex, corporate globalization has moved at an unprecedented rate to capture and expand market share, better serve global customers and improve corporate efficiencies. This global movement has produced a significant amount of economic benefit to the corporate business metrics. At the same time, globalization has opened a new realm of problems waiting to be solved. Problems like increased travel and information technology costs, communication barriers and additional capital and expenses associated with new or redundant engineering sites. It has been shown that the use of virtual prototypes along with well-defined and structured simulation models can improve the quality of the target software and overall system performance while reducing the products development time. This simulation capability allows the engineering team additional access to the product during the development process allowing problems to be found earlier in the development cycle while minimizing the rework effects on the physical hardware. [1] [3] If this same use of the virtual prototypes by the software development team could be utilized by systems and verification engineers, additional members of the development team could benefit as well, thus enhancing their ability to influence design changes prior to the manufacture of the physical product.

Today, systems benches and IT&V test environments involve a significant amount of complex modular test and simulation functionality capable of running static test cases as well as closed-loop simulation and dynamic systems and software tests. These environments contain a substantial amount of engineering and financial investment in order to provide for network capability, upgradeability and scalability, model-based operation, open architecture, plug-and-play operation, fault tolerance and most importantly, configuration control and scripted operation.

Likewise, substantial investments are also made by product teams and product lines to purchase physical equipment and to develop script and model re-use libraries. Assuming the investment is already in place for the physical simulator and infrastructure that supports the bench environment, it only makes sound engineering and business sense to take advantage of this groundwork.

For these reasons, Delphi continues to expand our existing systems and IT&V infrastructure so that it is capable of supporting its use with a virtual prototype. This will allow for complete transparency between the physical and virtual bench environments.

This paper will present an approach to expand the "Virtual Prototyping" technology from the typical software engineering user to the systems and IT&V engineering community. The ECU's targeted for the development of this project are a combination of a production intent electronic transmission control unit and an electronic engine control unit. The paper will first discuss the traditional product development and verification process along with the various types of bench simulators in use today. The typical methods used by systems and IT&V engineers for integration and verification of the product is also discussed. Next, a brief introduction and the associated benefits of virtual prototyping to product software will be presented. The virtual product development and verification process is then discussed. The paper will outline the method for constructing a "Virtual Test Bench" which includes a "Virtual ECU" and "Virtual Vehicle Simulator" while using the same traditional systems engineering methodologies. A comparison between the traditional development and the expanded virtual prototype development methods will then be reviewed. The paper will conclude with a summary of the benefits associated with using a virtual environment as an early enabler to product verification.

TRADITIONAL PRODUCT DEVELOPMENT AND VERIFICATION PROCESS

The traditional product development and verification process for an automotive ECU is best described by using the V diagram referenced in Figure 1. [4] The left side of the V diagram starts with the vehicle requirements. From the vehicle requirements, the system and subsystem is architected and defined in order to derive the ECU specification. At this point in the development flow, the ECU is designed and built as a physical prototype. While there is some simulation and analysis that takes place as the design traverses down the left side of the V in the development process, there is no full testing or verification of the system in the flow until after the physical product becomes available at the bottom of the V.

Once the physical product is made available, the verification process begins on the bottom, right side of the V at the ECU level, traverses up the right side of the V, and concludes at vehicle integration and testing. At each stage of the verification process, the results are fed back to the design process. The results are used to optimize the design in the next product cycle through the V. This feedback is represented by arrows between the left and right side of the V in Figure 1. Using this development approach, the ECU team is able to obtain feedback from the verification process in a reasonable amount of time. But the engineers at the system and sub-system level must wait even longer to obtain

feedback from the verification process in order to validate their design against the requirements. Essentially, in using the traditional engineering process, an engineering team must wait for a full product design, build and verification cycle in order to validate the physical product against the requirements. This is an expensive and time consuming process, but today it is proven and works well.

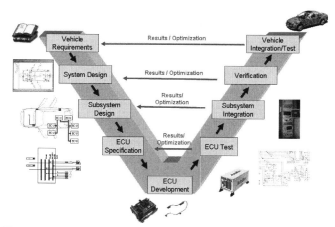

Figure 1 – The Traditional V Diagram

DEVELOPMENT BENCHES

Physical test benches are often comprised of customized static (manual) or dynamic simulators connected to the ECU under test via a wiring harness as depicted in Figure 2. Conventional testing using the manual simulators, as depicted in the top half of Figure 2, is slow and very dependent on the operator or engineer. Testing of the product is accomplished by an engineer or technician adhering to written test plans and manually documenting and reviewing the results for product compliance to the specification and/or requirements. For software verification, emulators are used to provide the visibility within the microcontroller. The emulators often require special silicon "bond-out" chips or Field Programmable Gate Array (FPGA) devices in order to make the internal registers visible to the software developer for monitoring and debugging purposes. As microcontrollers increase in complexity and frequency, extensive engineering development costs are required to accurately replicate the microcontroller. In some cases, the development of the emulator becomes technically unfeasible and the software engineer is left with only the "background debug mode" access to the silicon and registers.

The bottom half of Figure 2 depicts the more recent advancements in dynamic vehicle simulators which utilizes data acquisition boards to expedite and automate verification and testing of the target product. These computer controlled automated environments reduce the variability of the human operator, using a computer and automated script to drive the "switches" and the "loads" of the vehicle simulator. For closed loop control of the ECU, plant models are often incorporated into the host computer. Sometimes, these plant models can introduce complications resulting in potentially latent feedback to the ECU, but this can often be overcome through proper architecture of the environment. These newer methods come with higher equipment cost and can consume up-front engineering resources to develop interfaces and test scripts, but they offer higher test coverage, reuse, better debugging capabilities, repeatability and efficiency which more than offsets the development costs.

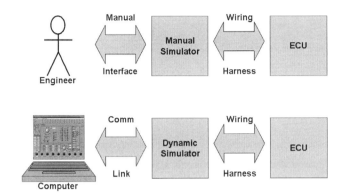

Figure 2 – Depiction of a Static (Manual) and Dynamic Simulator Set-up

A typical test bench using a dynamic simulator in the traditional product development flow is shown in Figure 3. The dynamic simulator is comprised of a microprocessor based hardware platform running plant models on a Real Time Operating System (RTOS) which simulates various vehicle sensor and load signals as well as other vehicle Input/Output (I/O) communication signals. As stated earlier, high fidelity plant models are used for the purpose of closed-loop real-time simulation. The simulation environment also contains an operational GUI to control and/or configure the vehicle signals as per the system requirements. The dynamic simulator is essential for complex and repetitive systems and IT&V activities.

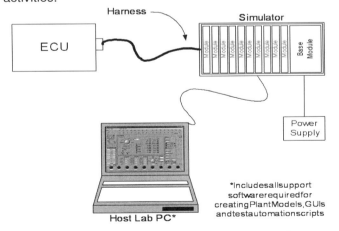

Figure 3 – Typical Dynamic Simulator Test Bench

ECU SYSTEM AND IT&V TESTING

Since simulation naturally lends itself to scripting and automation as a key method of improving productivity,

the dynamic simulator and its integration with a virtual prototype is the primary focus of this paper.

Various extensions are added to the dynamic simulator to meet the special needs of IT&V testing. A computer controlled relay panel is inserted between the ECU and the sensor signals as well as between the ECU and the loads. This provides manual or automated access to various I/O signals for monitoring and fault injection. For example, a solenoid output of the ECU can be subjected to the following "conditions" via relays:

- Connection of the solenoid output to the normal specified load
- Connection of the solenoid output to a short-to-battery (fault)
- Connection of the solenoid output to a short-to-ground (fault)
- Removal of any connection to the solenoid output in order to represent an open-circuit (fault)

The last three bullets are common vehicle wiring harness or component defects that are required to be detected and diagnosed by the ECU system. In addition, a diagnostic link such as CAN Communication Protocol (CCP) is commonly used by the scripts to determine if the ECU is able to properly detect each of the fault conditions within the required time. A typical script operation can be described as follows:

1. Connect the output to the solenoid
2. Initialization of the ECU
3. Read CCP to determine ECU is "fault free"
4. Open the relay connecting the solenoid (to induce open fault)
5. Wait the maximum specified time for detection of the open fault
6. Read CCP to determine that only the "open" was detected.

Software analysis is also commonly conducted on the system and/or IT&V bench. For example, the maximum allowed execution time of the ECU's fault detection software may be constrained to 400μs and specified to execute once every 10ms. The measurement starts at the beginning of the subroutine; however the global interrupts are commonly disabled to facilitate measurement of only the subroutine of interest. Repeated measurements for various fault conditions, transmission and Revolutions Per Minute (RPM) are taken to determine the minimum, mean and maximum latency for the subroutine. The measurement is often taken by a debugger script using the symbols from the program as provided by the IT&V symbol server. Typical symbols are memory map locations, variable sizes (by name) and entrance/exit points of the software.

In most cases, a limited number of hardware prototypes are made available to the IT&V team. It is not uncommon for prototype deliveries to be delayed due to combinations of software and hardware integration issues. It is also not unusual for the planned IT&V

evaluation schedules to become significantly compressed in order to meet program schedules or product deliveries. This implies that the IT&V engineer must develop tests without proper hardware or software in order to prepare for the intense integration and testing of the ECU when it is made available.

Today, most program development of the ECU is done by a collaboration of systems, software and IT&V groups distributed globally. The technical training along with the shipment of ECU's and dynamic simulators to the global development sites can be a significant problem causing addition delays in program schedules..

INTRODUCTION TO THE ECU VIRTUAL PROTOTYPING [1] [3]

As stated earlier, the complexity of the physical hardware and associated software development required to achieve the new generation of products in the automotive electronics market continues to increase. There continues to be a demand to increase "Time-to-Market" and "Design-to-Cost" capabilities within the engineering design process. The use of virtual prototypes in the engineering development process is one technology being used to help manage the product growth and development demands. The virtual prototype consists of a computer simulation model of a large part of a system or an entire product that can be used by many, if not all, of the engineering design team members.

The virtual prototype provides a new level of product visibility that is not typically available to many members of the engineering team using conventional development boards or actual product hardware. Since the entire product is created in a simulation environment, access to internal probe points within the design is almost unlimited. This allows for detailed concurrent debugging of the hardware and software in the target system.

The following bullets summarize the significant benefits that are gained through the use of virtual prototypes:

- Allows architecture, hardware, software and integration to start concurrently during product development in the absence of physical hardware.
- Enables an earlier start for engineering disciplines involved in the product design process which results in a shorter overall design cycle
- Enables a better understanding of hardware and software partitioning decisions and the determination of throughput considerations early in the development process
- Enables the use of functional models to determine microprocessor hardware configurations and architectures
- Enables the improved development of Application Specific Integrated Circuits (ASIC) architectures by improving the understanding and definition of requirements, functional performance improvements, and better documentation

- Enables application and system issues to be resolved in the silicon design phase resulting in fewer silicon revisions prior to production
- Promotes the efficiency of a global engineering workforce through its ability to be ported-to or accessed almost simultaneously by engineering resources located in corporate facilities worldwide

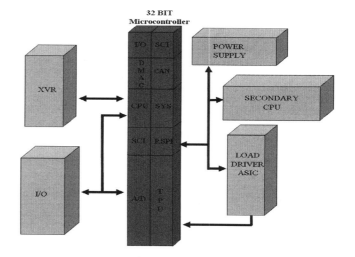

Figure 4 – Simple Block Diagram of the Electronic Control Module Virtual Prototype

Figure 4 shows a simple block diagram of a virtual prototype model, used in the development project, for a transmission control ECU. The system mechanization is comprised of a detailed functional model of the 32-bit microcontroller core along with its associated peripheral modules. The system also contains functional models of the power supply, custom transmission solenoid driver ASIC and a secondary microcontroller. This hardware model and simulation environment is capable of running the same target software in the same binary format as used in a physical product. The virtual prototype provides the software engineer the capability of simulating and debugging all aspects of the product software prior to, and independent of, the availability of the physical hardware. The Graphical User Interface (GUI) for this environment as seen by the software engineer is shown in Figure 5.

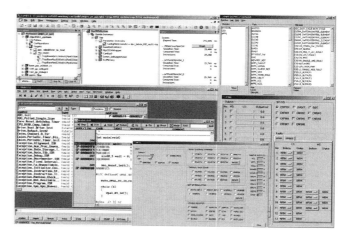

Figure 5 – GUI for the ECU Virtual Prototype

VIRTUAL PRODUCT DEVELOPMENT AND VERIFICATION PROCESS

To surmount the time pressure and high quality requirements of the vehicle manufacturers, automotive suppliers have been exploring and deploying model-based technologies as part of the product development process. Over the past decade there have been efforts to use the model-based or virtual technologies in requirements and ECU specifications, but the application of such technologies in product-level system testing and IT&V disciplines has been almost non-existent. Hence, a complete virtual development approach and integrated tool chain has been an uncharted territory until recently.

The intent of this paper is to only present enough information regarding the model-based V diagram, shown in Figure 6, to provide sufficient background on the topic. Only the key areas of development highlighted in the green ellipse will be discussed. Additional information on the model-based V diagram can be found in the reference section of this paper. [4]

As discussed earlier, when using the traditional engineering process an engineering team must wait for a full product design, build and verification cycle in order to validate the physical product against the requirements, which results in an expensive and time consuming process. Using a model based development flow enables immediate verification and feedback at each step of the design process as the development traverses down the left side of the V diagram.

Starting with the subsystem design phase of program development shown on the left hand side of the V diagram, the functionality of the subsystem can be modeled and simulated using common math-based design tools. Initially, at the subsystem level, modeling is done at a purely functional level and does not contain much or any detail about architecture or hardware partitioning. Partitioning of the hardware and software components typically starts within this phase. By using a model, iterative tradeoffs can be evaluated to optimize the partitioning.

Once an available model of the subsystem is created, the subsystem is capable of being simulated. The plant model is then constructed and integrated with the subsystem. At this point, verification of the subsystem design against the vehicle requirements can occur in simulation prior to moving to the next levels of ECU specification and development shown on the V diagram.

The details of the ECU specification, which include both hardware and software requirements are then finalized and the creation of the virtual prototype can commence. This leads directly into the ECU development phase without the need for a physical prototype. Since an introduction to virtual prototyping was discussed in some detail in the previous section, no further expansion on the topic will be presented.

Continuation up the right hand side of the V diagram from ECU test to subsystem integration can occur provided a test bench is available to exercise the virtual prototype. The development of an ECU test bench which is capable of exercising both the virtual product as well as the physical product is the motivation for this paper.

It is important to note that in all cases throughout the model-based V diagram, development and simulation of the model adds clarity and visual conceptualization of the system, subsystem and ECU specification, without making any commitments to an expensive hardware build cycle. This allows iterations and refinements of the design to the requirements at the lowest possible cost to the development team.

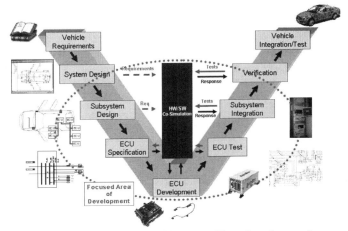

Figure 6 – Model-based V Diagram Showing Area of Focus

The ECU development for a virtual prototype is agnostic to auto-coded software, or traditional 'C' code. The virtual prototype executes the .elf or s-record load image as generated by the cross compiler and linker. The debugger is launched directly by the virtual prototype and provides breakpoints and tracing in a nearly identical manner to the bench. It should be noted that the emulator, bond out chips, and background debug mode have been replaced with a virtual interface to the CPU model of the virtual prototype.

The development bench starts as a "virtual test bench" with an emphasis on providing an operator interface and a scriptable test environment that will initially be used with the virtual prototype, then later deployed to the physical ECU. The test environment is the same dynamic simulator as used in the traditional process, but the back-end of it uses "virtual" I/O cards tied to the virtual prototype rather than physical cards in the test rack. The custom wiring harness is replaced by look-up tables in the scripts. The emphasis of the virtual test bench is to allow tests, first applied to the simulation, to be later applied or ported to the dynamic simulator and physical product. An important advantage of the virtual bench is that the environmental stimuli can be paused and resumed along with the product software from the debugger interface.

In moving up the right hand side of the V, peer reviews, static verification, unit testing, subsystem testing and software integration testing are nearly identical as the processes used in traditional development. The virtual prototype, however, allows for the coupling of math, state and UML based tools previously used in system development, linked to the host Personal Computer (PC) by direct target code communication. The tools facilitate software-in-the-loop and processor–in-the-loop simulations to test the software layers such as device drivers against the ECU hardware model, and then again with the physical hardware. Software development and testing can be carried out before the physical hardware is available. Essentially, in the virtual product approach, the entire white box testing of the software can be done without the physical hardware.

CREATION OF THE VIRTUAL TEST ENVIRONMENT

A feasible approach to creating a virtual system bench and IT&V test set-up is discussed and illustrated in this section. In this approach, a similar architecture to the physical bench is maintained. The physical ECU is replaced with a virtual prototype and the system bench set-up is moved completely into a simulation environment. Figure 7 shows a graphical comparison of an actual physical bench set-up to a virtual bench setup. Since the virtual bench is a direct adaptation of the physical bench, the same target binary/hex file test scripts used to run the automated tests on the physical bench can be used in simulation.

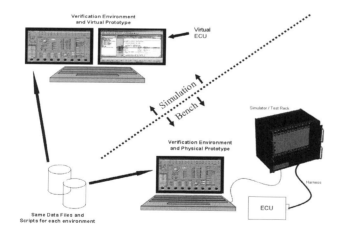

Figure 7 – Contrasting the Physical Bench to the Virtual Bench

USER INTERFACE

In order to ensure the software, system and IT&V engineers do not need to learn a new user interface for simulation, the same user interface provided by the physical test bench is provided in the virtual test bench. A communication interface is required to connect the GUI that is used for the dynamic simulator to the simulation platform that is executing the plant models. Generally, the GUI of the dynamic simulator is very slow

in comparison to the virtual ECU and plant models. Hence, the communication between this GUI and the virtual platform simulator is kept to a minimum to ensure that the simulation speeds of the test bench are not negatively affected.

PLANT MODEL

In the traditional test bench environment, the plant model forms the link between the GUI and the dynamic simulator hardware. The same architecture is preserved in the virtual test bench environment in order to maintain all of the functionality of the plant model. However, the plant model simulation platform is modified to interact with the virtual ECU platform. This is achieved by creating a bypass which allows the plant model to access and interact with the virtual prototype instead of the RTOS driven hardware. The modified model is then compiled and built for the PC environment. Further, with this modification, a common plant model can be used for the dynamic simulator and the physical bench by just changing a compile-time option. This approach ensures a smooth transition between the traditional and virtual benches with respect to the plant model.

SIMULATION PLATFORM

The simulation platform consists of several functional models representing various hardware interface modules on the dynamic simulator. They include a pulse generator module, pulse measure module, analog signal module, discrete input module, power moding module and a custom signal interface module. Each of these functional models mimics the behavior of various load and sensor circuits on a typical automotive system interacting with the ECU under development. Accordingly, these functional models are constructed to provide and receive the vehicle signals required by the ECU under the required timing constraints. The functional models of the interface cards simulate at the same rate as the virtual ECU in order to maintain timing synchronization. The architecture of the simulation platform is optimized to minimize and constrain the timing and communication with the user interface.

CONNECTING THE VERIFICATION TEST BENCH TO THE VIRTUAL ECU

The virtual ECU and the verification test bench typically share 25 to 50 connections between the two simulation environments. Each connection may represent voltage, resistance, current or even encoded frequency/duty cycle information. The information is exchanged using message packets sent via a network connection between the two independently executing simulation programs. A Delphi proprietary simulation framework manages the communications and scheduling between simulation programs and applications.

The virtual verification test bench is connected to a virtual powertrain ECU using an XML file format which contains the type of information represented in Table 1.

(Note: The table is for illustrative purpose only) The actual virtual ECU has many additional signals that are exchanged on a periodic basis which are not shown. The 4th row of the table, highlighted in *italic* font, shows a voltage signal called "BATTERY_ENABLE" on the test bench port which communicates changes to the equivalent port on the virtual ECU called VP.PWR_FLEX_CTRL1. The virtual ECU then applies the voltage to the VP.PWR_FLEX_CTRL1 port and advances the simulation. Any output change on the VP.BATTERY_STATUS port of the virtual ECU, shown on the 2nd row of the table and highlighted in **bold** font, is then communicated back to the test bench port named BATTERY_STATUS.

The overall simulation communication interval may be adjusted, to adapt to the type of signal exchange. A small change in the communication interval increases the fidelity of the simulation, while a large change increases the overall speed of the simulation.

Virtual ECU Name	Direction	Verification Test Bench Port Name	Units
VP.RS1_PRNDLA	output	RS1_PRNDLA	Volts
VP.BATTERY_STATUS	**output**	**BATTERY_STATUS**	**Volts**
VP.ROTARY_CONTROL	input	BRAKEPA	Volts
VP.PWR_FLEX_CTRL1	*input*	*BATTERY_ENABLE*	*Volts*
VP.TUPDNC6_R_VAR	input	TUPDNC6	Ohms
VP.PDL0_PWM0_LP	output	PDL_PWM0_LP	Hz (PWM)
VP.POM0.FREQ4	input	ETRSSP_FREQ	Hz (PWM)

Table 1 – XML File Format Example

A commercial vehicle simulator, capable of generating various vehicle signal conditions, is adapted for use with the virtual verification test bench. The virtual vehicle simulator is customized by Delphi through the creation of custom models and GUI. The simulator processes the information provided by the ECU outputs for status display or to compute further responses. For example, the vehicle simulator would generate crank and cam signals for different RPM conditions and processes the injector signals to calculate the injector duty cycle and frequency.

A commercial toolset and CPU core models are adapted to create the virtual powertrain ECU. Microprocessor peripheral models provided by the semiconductor suppliers along with ASIC models, GUI's and a simulation engine and back plane, developed by Delphi, complete the environment.

The end result is a front-end, virtual vehicle simulator user interface identical to the one used in the real bench. This gives the same look and feel for the software, systems and IT&V engineers as seen on the physical bench.

COMPARISON: TRADITIONAL TEST BENCH VS VIRTUAL TEST BENCH

Since the dynamic bench simulator is migrated directly from the physical bench to the virtual bench, a smooth and almost effortless bi-directional transition of tests cases can be realized. The virtual test bench shares the same user interface as the physical bench. There are no changes required in the product software or automated test scripts between the virtual and physical test benches.

The architecture of the dynamic simulator used on the physical test bench is split between a host PC running the GUI and second PC-like hardware set up running a RTOS which controls signal generation and measurement cards used to communicate with the ECU. This set-up gives an indication of the amount of computational power required at the PC level by the dynamic bench simulator.

Simulation speeds for the standalone virtual prototype are on the order of 1:5 to 1:20 of the physical hardware when executed on a typical single core PC or workstation. An additional CPU load introduced by running both the virtual prototype simulation and the dynamic bench simulator on a single core host machine slows down the simulation speed of the virtual prototype significantly. Architectural changes are made to the simulation environment to take advantage of a dual core engineering class workstation which allows the entire simulation of the virtual prototype and the dynamic virtual bench simulator to greatly improve. The speed of the virtual test bench simulation can also be enhanced by deploying two host PCs. As in the case of a dual core single PC approach, the GUI and plant model run on the first PC. The functional models of the vehicle simulators interface hardware and the virtual ECU run on the second PC. With this setup, the combined virtual test bench and virtual ECU simulation speeds are within 10% of the stand alone (without dynamic simulator) virtual ECU simulation speeds.

Performance limitations result from the introduction of additional user interfaces, scripts and test rack emulation as well as from the communication interface between the GUI and the virtual prototype. The communication interface abstraction levels between the two environments are required to be raised in order to gain simulation performance improvements. As an example, instead of communicating a Pulse Width Modulated signal (PWM) for a drive train sensor as discrete logical levels, only the frequency and duty cycle information are communicated between the user interface and the virtual prototype. The virtual prototype must then "reconstruct" the discrete signal prior to passing it into a timer channel of the microcontroller.

Both the virtual test bench and prototype provide the user increased visibility over the physical hardware. As an example, the internal registers of the solenoid driver ASIC are exposed as model "elements". This allows the user to confirm the current drive settings at a "white box", rather than "black box", level and eases the development of tests. In addition, the virtual prototype can be commanded to display the communication messages between the MCU and the various ASIC's, or to display the signal communications between the ECU and the user interface. Simulations are also easily instrumented with a "SPI Watcher" module that records the SPI transfers while a chip select is active. At the trailing edge of a chip select, the buffer is output with a simple print statement resulting in a display such as:

```
MCU SPI to EE: 0x03 0x00 0x3C 0x00 0x00
EE SPI to MCU: Hi-Z Hi-Z Hi-Z 0x9F 0xFF
```

Detailed parametric features such as the solenoid rise and fall time are not modeled in the virtual prototype to retain a reasonable level of simulation speed for software and systems development. While coupling a SPICE simulation into the virtual prototype is possible, the same analysis is often better accomplished via a decoupled VHDL-AMS simulation on a single solenoid or through a substantially downsized mixed signal simulation environment.

As product complexity continues to grow, software tool and computing platform suppliers will continue to be challenged to improve the simulation performance and maintain the viability of this technology. The continued deployment of dual and quad-core workstations will help counteract the product complexity, but careful partitioning of the overall simulation environment is required to take advantage of multi-core workstations.

Case Study One: Transmission Management System

For the first case study, the virtual environment is a relatively small and simple system using a 32-bit microcontroller along with a Delphi designed, high functionality, custom ASIC. The ASIC actively drives eight solenoid actuators which are located in or near to the vehicle transmission.

Typical independent testing for this type of a system is a combination of "black box" and "white box" testing. In both the physical and virtual bench environment, the vehicle simulator provides the sensor and actuator functionality. To date, the tests summarized in the following sub-sections have been completed successfully in both physical and virtual bench testing.

Discrete Outputs:

Basic commands, as shown in Table 2, are issued by the product ECU software to turn on and off a discrete output. The output is observed as an LED indicator on the vehicle simulator GUI.

Name	Variable	Set Value	Observed Output
SOLA	(TRANS_Discrete_Output_State.TRANS.SlndA)	1	12V:ACTIVE1
		0	0V:PASSIVE2
SOLB	(TRANS_Discrete_Output_State.TRANS.SlndB)	1	12V:ACTIVE1
		0	0V:PASSIVE2

Table 2 – Basic Discrete Output Command Run On The Physical And Virtual Bench

Discrete Inputs:

As shown in the data in Table 3, a switch on the vehicle simulator GUI is turned on. A corresponding variable in the software is then checked to have appropriate response value. It should be noted that on the virtual system, it is also possible to quickly verify the "ON" and "OFF" state voltage values and observe the software behavior.

Input Signal	Input Variable Name	Input Switch State	Software Variable Value
TRANSPS1	TransPreSw1	Open	0
		Closed	1
TRANSPS2	TransPreSw2	Open	0
		Closed	1
TRANSPS3	TransPreSw3	Open	0
		Closed	1
TRANSPS4	TransPreSw4	Open	0
		Closed	1

Table 3 - Basic Discrete Input Command Run On The Physical And Virtual Bench

Frequency Inputs

Tables 4A and B shows a PWM signal which is provided by the vehicle simulator according to the user specified frequency and duty cycle information. The appropriate software variable is checked to have the required rising edge counts. At this time, there are a few inaccuracies (data shown in Tables 4A and B with an asterisk) when comparing the physical and bench data which can be attributed to modeling errors and not the viability of the simulation environment.

Input Frequency (Hz) / DC	SPEED1_EDGE_COUNTS		
	Expected Value	HW Bench Value	Virtual Bench Value
20 / 50%	10	10	10
40 / 50%	20	20	16*
2000 / 25%	1000	1000	1106*

Table 4A - PWM Commands Run On The Physical And Virtual Bench

Input Frequency (Hz) / DC	SPEED2_EDGE_COUNTS		
	Expected Value	HW Bench Value	Virtual Bench Value
20 / 50%	10	10	10
100 / 50%	50	50	52*
1000 / 50%	500	502	564*

Table 4B - PWM Commands Run On The Physical And Virtual Bench

Case Study Two: Engine Management System

In the second case study, a virtual ECU of an engine management system is connected to the virtual test bench. This integration is more complex as the virtual ECU contains a 32-bit microcontroller along with several Delphi designed, high functionality, custom ASIC's and commercially available IC's. In this case, the number of communication signals between the virtual ECU and the test bench is greater than the system in Case Study One.

Some of the features and capabilities that result from the integration are as follows:

- Power control model to control battery and the ignition
- RPM signal generation model to generate crank and cam signals
- Pulse generating model to generate pulses at user defined frequency and duty cycle for transmission signals
- Pulse measuring module to measure the duty cycle and frequency of the injector signals.

Within this full environment to date, the user can power up the battery, set RPM and other signals in the virtual vehicle simulator and run the virtual ECU target software. A debugger provides full visibility of the target software within the virtual ECU.

It is extremely common for system or software engineers to create a set of test scripts which will execute regression tests on the physical bench. The virtual environment also accepts these same script commands. Hence, the same scripts that run on the physical bench can be run on virtual bench. To date, limited scripts have been run on both the physical and virtual bench. A simple sample script executed on both environments is illustrated below:

```
Log("Enabling Battery")
Set("Battery Enable",1)
Log("Crank...")
Set("Key Switch State",4)
Wait(500)
Log("Run.....")
Set("Key Switch State",3)
Log("Test script that update RPM value")
Set("RPM",3500)
```

BENEFITS TO DATE

As a direct result of the testing completed to date, it is evident that the virtual test bench will enable the systems and IT&V engineers to start test development and functional verification of the complete system as early as 6 months prior to the availability of the first hardware prototype. This is a significant benefit to the product development cycle time. Thus, the virtual development environment facilitates complete and thorough testing of the physical product on the first day of its availability to the engineering team.

Secondly, it appears that the overall cost of the virtual test environment can be lower than that of the physical test environment in equipment and product cost alone Adding to this benefit, the virtual bench can be deployed to any global engineering location in a matter of minutes. This offers minimal set-up, tear-down and removes the possibility of inadvertent bench disassembly.

Next, the combination of the virtual test environment with the virtual prototype is at its best to detect, analyze and fix hardware and software integration issues. (e.g. SPI bits transfer settings between the embedded microcontroller and SPI enabled ASIC's, A/D sampling rate settings for capturing vehicle sensor data, clock rate and PLL settings) Many of these details are painful to instrument and debug on the physical bench. From a systems viewpoint, any changes to loads, sensor circuits and crystal frequencies can be easily prototyped on the virtual development environment. These changes can then be analyzed and tested for feasibility from a functional point of view without building hardware. All of the above are significant benefits to the entire product development team.

Finally, once the initial investment is made to develop the simulation infrastructure and virtual model libraries for a given product type/family, subsequent virtual ECU and test environments in the same product family will require considerably less time than the creation of the first. This directly correlates to the level of reuse that is possible in such creations.

FUTURE INVESTIGATION AND IMPROVEMENTS

While the benefits seen to date are capable of supporting the use of a virtual development environment, there continues to be some aspects that need improvement. Specifically, as of this writing, product functionality that is subject to faster communication rates between the ECU and the test bench will require a small simulation interval resulting in slower overall simulation speeds. In addition, the interrupt latency and related checks cannot yet be suitably measured on the virtual bench. The software analysis probes inserted to the virtual ECU do not currently have mechanisms to exchange information with the system stimuli.

CONCLUSIONS

As a result of the project discussed in this paper, it is apparent that virtual product development environment is not only a significant advantage to the software engineering community, but it can provide substantial benefits to the systems and IT&V engineering communities. The key benefits of expanding the product level simulation environment to the systems and IT&V engineering disciplines can be summarized as follows:

- Enables bench and script development for systems and IT&V engineers early in the development process without hardware
- Enable systems and IT&V testing and feedback within the same V development cycle
- Supports rapid deployment to global development teams.
- Reduces the need for a large number of physical benches and prototypes

While this paper illustrates the many advantages of the virtual prototype and the virtual test bench, it is very important to point out that it does not replace the need for physical test benches or the need to verify the physical product. Instead, the virtual test bench supplements the physical test bench. A large portion of the functional verification of the software and system can be done in the virtual environment but the final system regression testing needs to be done on the physical bench and against the physical product. To fully achieve and exploit the potential of virtual product development, a complete library of models must be available. Continued engineering effort is also required to resolve the few technical shortcomings discussed in the previous section. In addition, a system and IT&V physical test bench strategy must be aligned with the virtual prototype and simulation test bench strategy to minimize the differences between the physical and virtual environments and costs of the entire development effort.

In summary, the use of a dynamic virtual vehicle simulator connected to a virtual prototype can shorten the "V" cycle and allows for earlier test development by systems and IT&V engineers. Such a strategy will ensure early and faster hardware/software integration, testing and feedback which ultimately improves engineering efficiency, product quality and development cycle times.

ACKNOWLEDGMENTS

The authors would like to acknowledge the significant engineering development efforts of Satheesh Mohan and Shashi Kiran in the Global Design Methodology and Automation Team located at Delphi's Technical Center India facility in Bangalore, India for their effort in bringing up the Virtual Test Bench for Engine Management and Transmission systems and Rick Nicholson in the Global Design Methodology and Automation Team located at Delphi Electronics and Safety World Headquarters in Kokomo, Indiana for his technical development of the proprietary simulation framework. In addition, the authors are appreciative of the technical support given by Mark McFarland and Don Geiselman of Delphi's Development Tools Group and Alan Soltis of Opal-RT Technologies for their application support for the vehicle simulator. Finally, the authors would like to thank Christine Winters for her dedication and time spent in editing and refinement of this paper.

REFERENCES

1. Winters, Frank; Mielenz, Carsten; Hellestrand, Graham, "Design Process Changes Enabling Rapid Development", Convergence 2004.
2. Lumpkin, Everett; Gabrick, Michael, "Hardware/ Software Design and Development Process", SAE World Congress 2006
3. Baron, Dr. Klaus U, "Virtual System Development - From Specification to Control Unit" Virtual Vehicle Creation 2007

CONTACT

Frank J Winters – Staff Engineering Manager, Global Design Methodologies, Delphi Electronics and Safety
Phone – 765.451.3009
E-mail – frank.j.winters@delphi.com

Everett R. Lumpkin – Senior Global Engineering Team Lead, Global Design Methodologies, Delphi Electronics and Safety
Phone – 765.451.3247
E-mail – everett.r.lumpkin@delphi.com

Chandrashekar M S – Technical Leader, Global Design Methodologies, Delphi Electronics and Safety
Phone – 9180.3077.7537
E-mail – ms.chandrashekar@delphi.com

Manjunath BC– Technical Leader, Global Design Methodologies, Delphi Electronics and Safety
Phone – 9180.3077.7531
E-mail – bc.manjunath@delphi.com

Verification and Validation According to IEC 61508:
A Workflow to Facilitate the Development of High-Integrity Applications

Mirko Conrad, Jonathan Friedman and Guido Sandmann
The MathWorks

ABSTRACT

Model-Based Design with production code generation has been extensively utilized throughout the automotive software engineering community because of its ability to address complexity, productivity, and quality challenges. With new applications such as lane departure warning or electromechanical steering, engineers have begun to consider Model-Based Design to develop embedded software for applications that need to comply with safety standards such as IEC 61508.

For in-vehicle applications, IEC 61508 is often considered state-of-the-art or generally accepted rules of technology (GART) for development of high-integrity software [6,11]. In order to demonstrate standards compliance, the objectives and recommendations outlined in IEC 61508-3 [8] must be mapped onto processes and tools for Model-Based Design.

This paper discusses a verification and validation workflow for developing in-vehicle software components which need to comply with IEC 61508-3 using Model-Based Design. It discusses tool support by using a Simulink based tool chain for Model-Based Design as an example.

INTRODUCTION

The increasing application of **embedded software** in commercial vehicles and passenger cars has resulted in a staggering complexity that has proven difficult to negotiate with conventional design approaches and processes.

More and more projects need to comply with external safety standards, because modern software-based electronic control units (ECUs) control or interact with critical systems such as brakes and steering. Examples include the electromechanical APA steering system [10] for the Volkswagen Tiguan [7].

Development approaches and processes play a decisive role in addressing the complexity, productivity, and quality challenges. Because of its ability to address these challenges, Model-Based Design has been extensively used throughout the automotive software engineering community. More recently, automotive OEMs and suppliers have begun to consider and adopt Model-Based Design for the development of embedded software for high-integrity applications that need to meet the IEC 61508 standard.

IEC 61508 is an international, industry-independent safety standard titled "Functional safety of electrical/electronic/programmable electronic safety-related systems". The seven parts of the standard (referred to as IEC 61508-1 to IEC 61508-7) were published between 1998 and 2000. IEC 61508-3 [8] "Software Requirements" is concerned with software development, verification, and validation.

Model-Based Design tools such as **Simulink**® and **Stateflow**® can be used throughout multiple phases of the software life cycle. Model-Based Design enables early simulation and testing of the functional behavior as well as automatic code generation for production systems.

SAE Int. J. Commer. Veh. | *Volume 2* | *Issue 2*

23

Production code generation within Model-Based Design has replaced manual coding in various vehicle domains and generated code is increasingly being deployed in high-integrity applications.

The successful use of Model-Based Design for high-integrity applications requires extra consideration and rigor because additional requirements imposed by IEC 61508 have to be met. The standard does not cover advanced software development, verification, and validation technologies, such as Model-Based Design and code generation. If high-integrity software is developed, using such technologies, objectives and recommendations of the standard must be mapped to the processes and tools used.

To ensure quality and functional safety, models and generated code should be subjected to a combination of quality assurance measures. Further, compliance with the IEC 61508 safety standard needs to be demonstrated.

This paper discusses how Model-Based Design and associated verification and validation techniques can be leveraged to develop automotive applications that comply with IEC 61508. It covers the mapping of selected IEC 61508-3 objectives onto Model-Based Design.

A WORKFLOW FOR VERIFICATION AND VALIDATION OF MODELS AND GENERATED CODE

IEC 61508-3, like other standards, calls for **application-specific verification and validation** regardless of the tool chain and the development process used. When using Model-Based Design with production code generation, application-specific verification and validation needs to answer the following questions:

- Does the model correctly implement its (textual) requirements?

- Does the object code to be deployed in the ECU correctly implement the model's behavior?

To facilitate a seamless use of Model-Based Design in the context of IEC 61508, a **reference workflow** that describes the verification and validation activities necessary for IEC 61508-3 compliance has been developed and audited by the TÜV SÜD certification authority.

Following the suggested division of the verification and validation task for applications developed using Model-Based Design and production code generation, the workflow would be divided into the following steps [1]:

1. Demonstrate that the model is correct and meets all requirements.

2. Show that the generated code is equivalent to the model.

The first step is **design verification**, which combines suitable verification and validation techniques at the model level. The second step is **code verification**.

DESIGN VERIFICATION — VERIFICATION AND VALIDATION AT THE MODEL LEVEL

The goal of design verification is to gain confidence in the model that is being used for production code generation. Design verification takes place at the model level, before the code is generated.

Following the spirit of IEC 61508-3, the design verification part comprises a combination of reviews, static analyses, and comprehensive functional testing activities at the model level [5]. These activities together help provide confidence that the design satisfies the associated requirements, does not contain unintended functionality and complies with defined modeling guidelines.

Requirements for design verification can be derived from IEC 61508-3 clauses 7.4.6 (code implementation), 7.4.7 (software module testing), and 7.4.8 (software integration testing).

Reviews and Static Analyses at the Model Level

Model subsystems considered as modules (model components) should be reviewed. If feasible, manual model reviews should be supported by automated static analyses of the model.

Modeling guidelines should be used and adherence with the guidelines should to be assessed. Modeling constructs that are not suited or not recommended for production code generation should not be used.

Tool Support - Model reviews can be facilitated by using reports or interactive renditions (Web views) generated with Simulink Report Generator. Adherence to a modeling guideline can be partially enforced by using predefined or customized checks in Model Advisor [3], including MathWorks Automotive Advisory Board (MAAB) and IEC 61508 checks.

Module and Integration Testing at the Model Level

Model components should be functionally tested using systematically derived test vectors. The objective of module testing is to demonstrate that each model component performs its intended function and does not perform any unintended functions. The testing techniques mentioned in IEC 61508-3 table B.2 (dynamic analysis and testing) can be used to derive test vectors for the model.

24

SAE Int. J. Commer. Veh. | *Volume 2* | *Issue 2*

As module testing is completed, module integration testing should be performed with predefined test vectors, i.e., the model integration stages should be tested in accordance with the specified integration tests. These tests should show that all model modules and model subsystems interact correctly to perform their intended function and do not perform unintended functions.

A more detailed discussion on testing at the model level can be found in [4].

Tool Support - Simulink and Simulink Verification and Validation support various facets of model testing.

CODE VERIFICATION — VERIFICATION AND VALIDATION AT THE CODE LEVEL

The starting point for code verification is a **golden reference model**, i.e., a sufficiently validated and verified, well-formed model that implements the requirements and does not contain any unintended functionality.

We utilize **translation validation through systematic testing (translation testing)** to demonstrate that the execution semantics of the model is being preserved during code generation, compilation, and linking. Translation testing in this context means that each translation (i.e. a run of the code generation tool chain) is augmented with an **equivalence testing** phase that verifies that the executable derived from the generated C code (object code) correctly implements the golden model. Equivalence testing is combined with additional measures to demonstrate the **absence of unintended functionality**.

Figure 1 gives an overview of the proposed verification process for generated code.

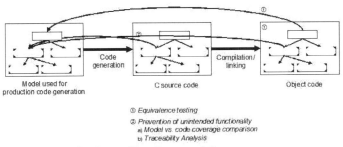

① Equivalence testing
② Prevention of unintended functionality
  a) Model vs. code coverage comparison
  b) Traceability Analysis

Figure 1: Code verification workflow.

Equivalence Testing

Numerical equivalence testing refers to the stimulation of both the model used for code generation and the object code derived from it through code generation and compilation with identical test vectors. This activity (also known as **comparative testing** or **back-to-back**

testing) constitutes a core part of the code verification process.

The validity of the translation process—whether the semantics of the model have been preserved during code generation, compilation, and linking—is determined by comparing the **system reactions**, i.e. **result vectors**, of the model and the generated code resulting from stimulation with identical timed test vectors $i(t)$. More precisely, the simulation results $o_{SIM}(t)$ of the model used for production code generation are compared with the execution results of the generated and compiled production code $o_{CODE}(t)$ (Figure 2).

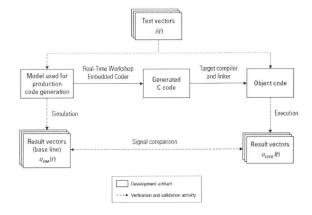

Figure 2: Numerical equivalence testing.

Testing for numerical equivalence is unique in that the expected outputs for the test vectors do not have to be provided [1]. This makes equivalence testing well suited to automation.

The following subsections provide an overview on the equivalence testing procedure. In-depth discussions of equivalence testing procedures are available in [12, 13, 14].

Equivalence Test Vector Generation - A valid translation requires that the execution of the object code exhibits the same observable effects as the simulation of the model for any given set of test vectors. Since complete testing is impossible for complexity reasons, stimuli (test vectors) shall be sufficient to cover the different structural parts of the model.

To assess the model coverage achieved, some **test coverage metric** shall be visible at Safety Integrity Level (SIL) 2 and above [15]. The extent and scope of structural model coverage needs to be increased for the higher SILs. Suggested metrics can be found in [9].

Tool Support - Model coverage analysis can be performed by using the Model Coverage Tool in Simulink Verification and Validation.

Figure 3 illustrates equivalence testing of individual and integrated modules.1

Test vectors resulting from requirements-based testing at the model level can be reused for equivalence testing.

If the coverage achieved with the existing test vectors is not sufficient, additional test vectors should be created manually or test vector generation tools such as

Signal Comparison - After test execution, the result vectors (simulation results) $o_{SIM}(t)$ of the model should be compared with the execution results $o_{CODE}(t)$ of the generated code. The simulation results of the model are used as the baseline. Even if there is a correct translation of a Simulink / Stateflow model into C code, one cannot always expect identical behavior (equality). Possible reasons include limited precision of floating point numbers, quantization effects when using fixed

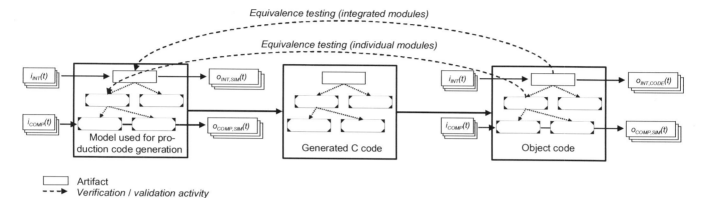

Figure 3: Equivalence testing of individual and integrated modules.

Simulink Design Verifier. In practice, the set of test vectors can be iteratively extended using model coverage analysis until the mandated level of model coverage has been achieved.

If full coverage for the selected metric(s) cannot be achieved, the uncovered parts should be assessed and justification for uncovered parts provided.

Equivalence Test Execution - The test vectors for equivalence testing shall be used to stimulate both the model used for production code generation and the executable derived from the generated code. The resulting object code shall be tested in an execution environment that corresponds as far as possible to the target environment in which the code will be deployed.

The resulting object code can be either executed on the target processor or on a target-like processor, e.g., by means of a processor-in-the-loop (PIL) simulation, or simulated by means of an instruction set simulator (ISS) for the target processor. When feasible, PIL verification is preferred over ISS verification.

If the execution of the resulting object code is not carried out in the target environment, differences between the testing environment and the target environment should be analyzed in order to make sure that they do not adversely alter the results.

point math, and differences between compilers. So the definition of numerical correctness has to be based on sufficiently similar behavior (sufficient similarity).

A suitable signal comparison algorithm should be selected that is able to tolerate differences between the result vectors representing the system responses $o_{SIM}(t)$, and $o_{CODE}(t)$. There is a broad variety of potential comparison algorithms, ranging from simple algorithms, like absolute difference, to more elaborated ones, such as the difference matrix method [16].

Two result vectors are sufficiently similar if their difference with respect to a given comparison algorithm is less than or equal to a given threshold. The selection of the comparison algorithm and the definition of the threshold value depend on the application under consideration and need to be documented.

Tool Support - Simulink Fixed Point allows performing bit-true simulations of model portions implemented using fixed point math to observe the effects of limited range and precision on. When used with Real-Time Workshop Embedded Coder, Simulink Fixed Point enables pure integer C code to be generated from these model portions. The generated code is in bit-true agreement with the model used for production code generation.

A model and the code generated from it are regarded to be functionally equivalent if the simulation of the model and the execution of the executable derived from the generated code lead to sufficiently similar result vectors

---

1 The functions $i_{COMP}(t)$ and $o_{COMP}(t)$ refer to the test vectors and result vectors for model components respectively; $i_{INT}(t)$ and $o_{INT}(t)$ refer to the test vectors and result vectors for the integration stages, respectively.

26

SAE Int. J. Commer. Veh. | Volume 2 | Issue 2

if both are stimulated with identical test vectors that satisfy the selected structural coverage criterion.

Tool Support - Comparison algorithms can be implemented in a general-purpose programming or scripting language such as MATLAB®.

<u>Prevention of Unintended Functionality</u>

The second activity in the code verification process is to demonstrate that the generated C code does not perform any unintended function i.e. the code does not contain functionality not specified by the model.

According to the workflow, two alternative techniques are available to achieve this objective. They serve the purpose of demonstrating structural equivalence between the model and the source code.

By using at least one of the following measures it can be demonstrated that the generated C code does not perform any unintended function:

- Model versus code coverage comparison
- Traceability analysis

Model Versus Code Coverage Comparison - Model and code coverage are measured during equivalence testing and compared against each other. Discrepancies with respect to comparable coverage metrics should be assessed.

To be meaningful, structural coverage metrics comparable with each other should be used on the model and code level, respectively. According to [2], decision coverage at the model level and branch coverage (C1) at the code level (which is also sometimes termed decision coverage) can be used in combination.

Discrepancies between model and code coverage with respect to comparable metrics shall be assessed. If the code coverage achieved is less than the model coverage, unintended functionality could have been introduced.

Tool Support - Code coverage information can be derived from the integrated development environment (IDE) or by applying a standalone code coverage tool to the generated source code.

Traceability Analysis - A **traceability analysis** of the generated C source code can be performed to ensure that all parts of this code can be traced back to the model used for production code generation. In this case the generated code is subjected to a **limited review** that exclusively focuses on traceability aspects. Nontraceable code shall be analyzed.

Tool Support - Automatically generated Simulink block comments can be used to generate tracing information into the generated code. The traceability report generated by the Real-Time Workshop Embedded Coder code generator report helps to provide a complete mapping between model blocks and the generated code. Code-to-block highlighting generates hyperlinks within the displayed source code to view the blocks or subsystems from which the code was generated. Block-to-code highlighting allows for any block in the model to identify the resulting generated code.

SUMMARY AND CONCLUSION

Currently, IEC 61508-3 is a relevant standard with respect to software development for high-integrity in-vehicle applications. It defines requirements and constraints for software development and quality assurance processes. These requirements apply to both Model-Based Design and traditional software development. However, implementing these requirements within Model-Based Design requires special consideration and creates specific challenges.

In this paper, the authors discussed workflows for Model-Based Design and tool considerations that address the objectives of IEC 61508-3.

The proposed workflow for verification and validation of models and generated code can be viewed as instantiation of the verification and validation requirements outlined in IEC 61508-3. This instantiation exploits the advantages of production code generation as an integral part of Model-Based Design.

A more detailed version of this workflow has been discussed with TÜV SÜD and was a foundation for the in-context certification of Real-Time Workshop Embedded Coder to IEC 61508. This detailed workflow description, a compliance demonstration template, and corresponding tool certification evidence (certificate and certification report) are part of IEC Certification Kit [9]. The TÜV SÜD certification for the Real-Time Workshop Embedded Coder code generator is based on satisfying the IEC 61508 requirement to use certified translators for SIL 2 and above (cf. IEC 61508-3 clause 7.4.4.3a and table A.3).[2]

REFERENCES

1. Aldrich, W., Coverage Analysis for Model-Based Design Tools, TCS 2001.
2. Baresel, A., Conrad, M.,Sadeghipour, S., Wegener, J., The Interplay Between Model Coverage and Code Coverage, 11. European Int. Conf. on Software Testing, Analysis and Review (EuroSTAR '03), Amsterdam, Netherlands, 2003.

---

[2] Note that Real-Time Workshop Embedded Coder was not developed using an IEC 61508 certified process.

SAE Int. J. Commer. Veh. | Volume 2 | Issue 2

27

3. Begic, G., Checking Modeling Standards Implementation, The MathWorks News & Notes, June 2007.

4. Conrad, M., Fey, I.: Testing Automotive Control Software. In: Navet, N., Simonot-Lion, F. (Eds): Automotive Embedded Systems Handbook, CRC Press, 2009.

5. Conrad, M., Model-Based Design for IEC 61508: Towards Translation Validation of Generated Code, Proc. Workshop Automotive Software Engineering: Forschung, Lehre, Industrielle Praxis, colocated with Software Engineering 2008, Munich, February 2008.

6. Faller, R.: The Evolution of European Safety Standards. exida.com, 2002.

7. Fey, I., Müller, J., Conrad, M., Model-Based Design for Safety-Related Applications, Proc. Convergence 2008, Detroit, MI, USA, Oct. 2008.

8. IEC 61508-3:1998, Int. Standard Functional Safety of Electrical/Electronic/ Programmable Electronic Safety-Related Systems - Part 3: Software Requirements, First edition, 1998.

9. IEC 61508 Certification Kit. The MathWorks, Inc., www.mathworks.com/products/iec-61508.

10. Jablonski, T., Schumann, H., Busse, C., Haussmann, H., Hallmann, U., Dreyer, D., Schöttler, F.: Die neue elektromechanische Lenkung APA-BS, ATZelektronik 01/2008 Vol. 3 (2008) 01, pp. 30-35.

11. Lovric, T.: Sicherheitsanforderungen an Fahrerassistenzsysteme. 2. Sachverständigentag (SVT 06), 2006.

12. Stürmer, I.; Conrad, M.: Test Suite Design for Code Generation Tools. 18. IEEE Int. Conf. on Automated Software Engineering (ASE '03), Montreal, Canada, 2003.

13. Stürmer, I.; Conrad, M.: Ein Testverfahren für optimierende Codegeneratoren. Inform. Forsch. Entwickl. 19(4): 213-223 (2005).

14. Stürmer, I.; Conrad, M.; Dörr, H.; Pepper, P.: Systematic Testing of Model-Based Code Generators. IEEE Transactions on Software Engineering, Sep 2007, pp. 622-634.

15. Smith, D. J.; Simpson, K. G. L.: Functional Safety. Elsevier, 2005.

16. Conrad, M.; Sadeghipour, S.; Wiesbrock, H-W.: Automatic Evaluation of ECU Software Tests. SAE 2005 Transactions, Journal of Passenger Cars - Mechanical Systems, March 2006

CONTACT

Dr. Mirko Conrad
Development Manager - Simulink Certification and Standards; The MathWorks, Inc.
Mirko.Conrad@mathworks.com

Mirko Conrad manages MathWorks certification and standards team. He oversaw the certification of Real-Time Workshop Embedded Coder to IEC 61508 and the development of IEC Certification Kit. He started his professional career in the automotive industry in 1995 and joined The MathWorks in 2006.

Mirko holds a Ph.D. in engineering (Dr.-Ing.) and an M.Sc. in computer studies (Dipl.-Inform.) from Technical University Berlin, Germany. He is also a visiting lecturer at Humboldt University in Berlin. His publication record includes more than 60 papers on automotive software engineering, Model-Based Design and safety-related software. He is a member of the Special Interest Group for Automotive Software Engineering in the German Computer Society (GI-ASE) and was a member of the ISO 26262 sub-working group on software.

Dr. Jonathan Friedman
Manager - Aerospace & Defense and Automotive Industry Marketing; The MathWorks, Inc.
Jon.Friedman@mathworks.com

Guido Sandmann
Automotive Marketing Manager, EMEA; The MathWorks GmbH
Guido.Sandmann@mathworks.de

28

SAE Int. J. Commer. Veh. | Volume 2 | Issue 2

Hardware/Software Design and Development Process

Everett Lumpkin and Michael Gabrick
Delphi Corporation, Electronics and Safety Division

ABSTRACT

In today's extremely competitive, technically challenging and cost sensitive market place, the design process must be reviewed to shorten the design time and remove the expenses associated with building engineering development controllers. Embedded system software is steadily increasing in size and sophistication, as well as the complexity of the hardware and environmental interfaces which adversely impacts the Embedded Control Unit (ECU) time-to-market and quality. To remedy the increasing complexity, the trade-offs between software and hardware can now be investigated via concurrent development early in the design cycle. Virtual systems provide a method to evaluate the interaction of microprocessors, memories, peripheral devices as well as software prior to hardware realization. This paper discusses and demonstrates the productivity improvements that can be realized using new virtual system design tools and methodologies to enhance the traditional design process.

INTRODUCTION

Process and technology advancements in the semiconductor industry have helped to revolutionize automotive and consumer electronics. As Moore's Law predicted, the increase in complexity and operating frequencies of today's integrated circuits have enabled the creation of system applications once thought to be impossible. End systems such as camera cell phones, automotive infotainment systems, advanced powertrain controllers and handheld personal computers have been realized as a result.

In addition to the increases in process technology, the Electronic Design Automation (EDA) industry has helped to transform the way semiconductor integrated circuits (IC) and subsequent software applications are designed and verified. This transformation has occurred in the form of design abstraction, where the implementation continues to be performed at higher levels through the innovation of design automation tools.

An example of this trend is the evolution of software development from the early days of machine-level programming to the C++ and Java software written today. The creation of the assembler allowed the programmer to move a level above machine language, which increased the efficiency of code generation and documentation, but still tied the programmer to the underlying hardware architecture. Likewise, the dawn of C / C++ compilers, debuggers and linkers helped to move the abstraction layer further away from the underlying hardware, making the software completely platform independent, easier to read, easier to debug and more efficient to manage.

However, a shift to higher levels of software abstraction has not translated to a reduction in complexity or human resources. On the contrary, as integrated systems have become more feature rich, the complexity of the operating system and corresponding applications have increased rapidly, as have the costs associated with the software implementation and verification activities. Certainly the advancements in embedded software tools such as static code checkers, debuggers and hardware emulators have helped to solve some of the software verification problems, but software verification activities have become more time and resource consuming than the actual software creation. Time-to-market constraints have pushed software verification activities to the system-level, and led to a greater demand for production hardware to be made available earlier in the software development flow.

As with the software case, the semiconductor design community has made a very similar transformation in their design and verification strategies sparked by advances in the EDA community. Designs that were once implemented completely at the transistor level migrated to the gate-level implementation through the development of schematic capture tools. The creation of hardware description languages such as Verilog and VHDL and the corresponding compilers, simulators and synthesis tools allowed hardware designers to move away from the gate-level implementation to the register transfer level (RTL). The EDA community is now promoting even higher levels of abstraction, often under the banner of electronic system level design (ESL) [6]. Again, this represented a fundamental change in design abstraction, which allowed the designers to think in terms of overall functionality instead of the configuration of gates needed to implement the desired functionality.

As Application Specific Integrated Circuit (ASIC) design complexities have grown and the process geometry continued to shrink, the manufacturing and NRE costs for silicon has increased rapidly. For example, the cost for silicon mask sets range from $50,000 for a simple ASIC to greater than $1,000,000 for an advanced microprocessor or microcontroller [4]. The high costs associated with ASICs underscores the motivation of the hardware community to insure that the intended functionality is implemented correctly prior to taking a design to silicon. The EDA industry has helped this cause by providing sophisticated verification tools that prove the high-level design and the silicon implementation will function equivalently. However, even with these tools available, more than ½ of all IC and ASIC designs require a re-spin of silicon, where 70% of the re-spins are due to logic or functional errors that verification efforts should have caught [5]. With the huge investment required for each re-spin, system level verification is becoming a focus of the overall hardware verification strategies.

Although we have seen significant advancements in the processes of hardware and software design during the past two decades, surprisingly, there have been little advancements made at the system level. Today's system process consists of the paper study of the proposed hardware architecture, required functionality, microprocessor throughput, memory configuration, and the potential hardware migration paths. The process has remained relatively unchanged. Furthermore, the software implementation is typically held off until hardware prototype units are created, placing the software developers and system verification teams at a disadvantage. This current approach has many drawbacks including: slow adaptation to changes in customer requirements, drawn out hardware and software integration, limitations in system debugging, and difficulties meeting the time-to-market constraints.

This paper presents a new approach to system-level design through the creation of a virtual system, which allows for an early analysis of hardware and software interaction while removing many of the drawbacks plaguing traditional system development. This paper also presents a virtual automotive air-bag system implementation and explores the benefits of virtual system development.

PARADIGM SHIFT

The motivation for system level design and analysis is to significantly improve productivity through a paradigm shift that allows hardware and software to be designed concurrently rather than serially. Productivity is thus enhanced by the elimination of re-work, increased quality of the final system, improved verification, and shorter time-to-market.

As design trends continue to move to higher levels of abstraction, more emphasis will continue to be placed in verification activities at both the component and system level. The creation of a "virtual" system using accurate models of the hardware provides engineers with the following benefits: an architectural exploration of hardware and software functions, the creation of flexible prototype hardware, more accurate analysis of throughput and portability, software development earlier in the cycle, and rapid debugging through the instrumentation of the virtual hardware.

One of the primary advantages of a virtual system implementation is architecture exploration, which is the process of determining an optimum solution to a design space problem. Take for example the two-dimensional architectural space shown in Figure 1. The two design parameters shown (typically there are many design parameters) are power consumption and clock speed, with the ideal solution illustrated by the center of the target. In this simplified example, the yellow cross illustrates a hardware prototype ECU that exceeds the ideals for power consumption and fails to meet the clock speed ideals. Because of the time-to-market constraint, the system architecture continues to be based upon the initial hardware prototype without adequate exploration of alternative choices. The end result of a hardware based development process is a suboptimum product that may miss the design targets.

Conversely the green crosses show an alternative path to the optimum design solution. Several virtual systems are assembled and tested in the same time-to-market window. The final virtual system is on-target for the design parameters, and the resulting work products from the virtual system are quickly converted into the physical product. Models of the system are initially created at a high level of abstraction, and, through the model-based methodology, are driven to a full virtual implementation, then to an actual product.

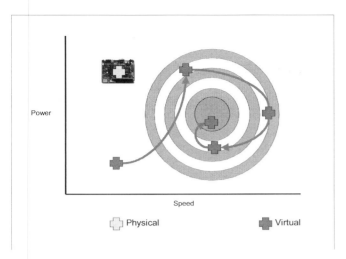

Figure 1 – Exploration of the design space using physical and virtual prototypes

The model-based methodology approach proposes the use of an architecture exploration tool to facilitate the rapid exploration of various CPUs, memories, and peripherals (system architecture). The system architecture is shown in the upper right of Figure 2 [7]. The upper left of Figure 2 shows early revisions of the functional models of the system, known as system behavior. The system behavior and system architecture are combined, and an optimal solution is achieved by iteratively comparing the performance of each partitioned alternative. The objective is to evaluate the various architectural and partitioning alternatives to gain a first order approximation of the optimum design.

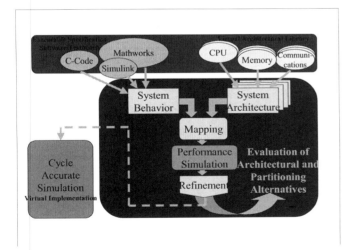

Figure 2 – Architecture exploration process

Some of the proposed EDA tools for architecture exploration [1] offer the ability to model microcontroller architectures in only a few weeks duration. The simulations for architectural simulation may be only 80% accurate, but that is believed to be good enough to make the first order choices of microcontroller, timer architecture and memory usages. Our evaluations of the architectural exploration tools, however, indicate that the industry has not yet focused on solving microcontroller selection in a general way. As this paper presents, there is a tradeoff between simulation accuracy, cost of the modeling effort, and the time to model a new architecture. We are awaiting further tool maturation before expanding beyond paper evaluations of architecture exploration.

The industry trend is toward building libraries of very accurate architecture models that execute the embedded software directly. A sufficient library of these detailed models can then be used to evaluate system architectures and also used for the final software development.

We have focused our efforts on the lower left box of Figure 2; the highly accurate and fast simulation of a virtual system known as cycle-accurate simulation. This portion of the co-design market has matured to the point that it is now feasible to simulate a moderate sized ECU, such as an air-bag deployment module, at about $1/15^{th}$ the speed of the actual hardware. As additional models become available, these cycle-accurate simulations will be capable of solving much of the architectural exploration space as well.

Detailed and highly accurate simulation encourages evaluation of the system behavior on a proposed architecture. The models, as well as the target code, can be adapted to determine the optimum design solution. Full target software simulation is supported using a 32-bit CPU Virtual Processor Model (VPM), the microcontroller peripheral models, system ASIC models, and the environmental stimuli. The VPM is able to load and simulate the same executable image as used in traditional bench development. The models may be exported from CoMET to the lower cost METeor integrated development environment to provide a fixed platform (models cannot be changed) virtual product simulation.

SIMULATION SPEED AND ACCURACY

Before we begin discussing the system, it is important to present some background material that highlights the levels of modeling abstraction and how they are best applied.

In general, total simulation speed decreases as detail is added to the model. Functional models execute orders of magnitude faster than the models used during co-verification. The levels of abstraction used in product modeling include [3]:

- Functional Simulation – Very little to no timing accuracy of the real hardware.
- Timed Functional Simulation – Contains estimated execution time of individual modules.
- Cycle-Approximate – Timed functional simulation techniques applied to instruction set and cycle-accurate simulation.
- Instruction Set Simulation – Cross-compiled code executed on model of target CPU.
- Cycle-Accurate – Simulation is very similar (or identical) to hardware behavior.

The remainder of this section discusses each of these abstraction levels in greater detail.

FUNCTIONAL SIMULATION – Pure functional simulation is often faster than the actual product implementation due to the speed of the host platform compared to the embedded system. The behavior of the individual modules is modeled in a sequential manner. The system behavior model is then compiled and executed directly on the host platform, providing excellent simulation speed. This level of modeling is extremely useful for algorithm development. An example of functional simulation is a Matlab Simulink algorithm. Functional simulation is typically used to discover and refine the requirements of the system.

TIMED FUNCTIONAL SIMULATION – Adding timing constraints or "timing budgets" becomes a very powerful addition to the functional simulation model. Each module is assigned a timing constraint based on its estimated communication throughput, hardware complexity or software instruction throughput capability. The simulation kernel for this level of simulation has the capability to denote passage of time and detect overflow (over-budget) conditions. The accuracy of this simulation is near that of the final hardware, yet the simulation proceeds without continual context switching. This model has a reasonable tradeoff of simulation speed vs. accuracy for use in partitioning studies as well as software development. Timed functional simulation is used to evaluate system architecture and it's ability to support the desired system behavior.

Performance modeling (co-design) of the system is best conducted with this level of simulation. As timing estimates are added to the various modules, they are modeled together to insure that the timing assumptions do not violate the overall system performance budget.

INSTRUCTION SET SIMULATION – Instruction set simulators (ISS) use cross-compiled target code to automatically derive timing information. This information is used to interface the application software with cycle-accurate simulations of the hardware. The simulation speeds (when coupled with traditional Verilog simulations) are typically too slow for use in actual software development, as just a few seconds of product operation may require hours or days of simulation.

Through careful attention to detail, instruction set simulations can be made to execute at speeds acceptable for software development (10-1000 million instructions per second). Instruction set simulation is best used for the debugging and optimization of embedded code.

CYCLE ACCURATE – Cycle-accurate simulation means that the hardware or software is executed in a manner identical to the actual hardware. The logic associated with a clock cycle is executed every cycle, such as timers, counters, etc. This level of modeling often requires a simulation kernel that supports concurrent operations, such as a Verilog or VHDL simulator. The models are timing accurate (typically within tens of nanoseconds) and simulation will be slow. Cycle-accurate simulation is used for hardware development. We often use the term "cycle-accurate" to loosely describe the desired accuracy of a system level simulation, but the more appropriate term is cycle-approximate".

CYCLE APPROXIMATE – Cycle-approximate simulation attempts to combine the best of instruction set, timed functional and cycle-accurate simulations to achieve a simulation with the fidelity that appears to be cycle-accurate, yet approaches simulation speeds similar to timed functional simulation. The techniques used to increase simulation speed include:

- Execution of "pure software" in bursts until a specific latency or hardware register access is met
- Timed event delays rather than counting cycles
- Reduction of decision (if, switch) statements in the simulation models, often via the use of function pointers and state machines
- Execution of update logic only when necessary, such as when a communication path is read instead of every time it is written
- Condensing serial communication (bits) into a parallel transfer of the data at the appropriate time
- Removal of clocking signals and schemes
- Aggregation of discrete signals into buses or ports

For effective virtual system simulation, the primary modeling level is cycle-approximate, but uses the above techniques to replicate the accuracy of cycle-accurate modeling. The accuracy is effectively the same as cycle-accurate simulation (typically tens of nanoseconds). Cycle-approximate simulations are used to develop software against full models of the peripherals and ASICs, and can also be used for some architecture exploration activities. The virtual system described in this paper uses a combination of timed functional, cycle-accurate and cycle-approximate simulation.

AIR-BAG VIRTUAL SYSTEM

The air-bag virtual system has the equivalent functionality of a recent production air-bag ECU. The module consists of a 32-bit microcontroller coupled with 7 ASIC devices as shown in the right side of Figure 3.

The air-bag system software reads acceleration data directly from accelerometers within the ECU, and also receives acceleration/severity data from remote sensors that communicate to the ECU. The algorithms make a deployment decision based upon the acceleration "crash waveform" data as well as vehicle state information such as seat belt engagement, seat position and occupant detection. There are 5 different serial interfaces used by the target software. The software continuously communicates with the ASIC devices that control deployment, and executes various diagnostics on the deployment loops. In the event of a severe crash, the 32-bit microcontroller and a simple "safe-ing" microcontroller must agree and command the air-bag deployment loop via the deployment ASIC. The ECU also contains a serial EEPROM device that is used to record operating statistics such as ignition cycles as well as "event" information during deployment or near-deployment.

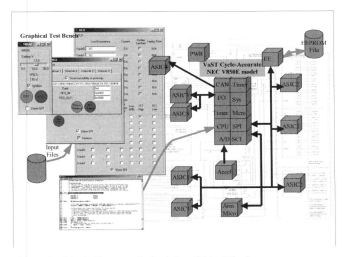

Figure 3 – Block diagram of an air-bag Virtual System

The 32-bit RISC based microcontroller includes various peripheral functions: less than 100 discrete I/O pins, DMA controller, interrupt controller, clock generator, 16-bit timers, watchdog timer, UART, serial peripheral interface, A/D and a CAN interface. As such it represents the medium complexity microcontroller commonly used in restraint and body functions of an automobile.

In the virtual system the microcontroller model loads the target executable image and symbol tables. The executable image is the same as used for bench development; even using the same compiler optimization flags and compile options. The microcontroller peripherals such as the A/D, timers, UART, SPI, and IO pins are also part of the simulation model, and are written as 'C' timed behavioral-level functional models. The models are interconnected using a subset of Verilog to describe the coherency and hierarchal portions of the design. The peripheral models were supplied in object form (no source code). Each model is accompanied by documentation showing interconnections; register map, embedded warning messages, and configuration options. Full interrupt capability of the peripheral models is supported, and a peripheral interrupt will cause the CPU model to take the exception and begin execution of the interrupt handler code, just as the hardware would behave. The models appear to be cycle-accurate, however some abstractions are applied to the behavioral models. For example a timer counter register may only be updated when the target software accesses the timer or when an interrupt occurs.

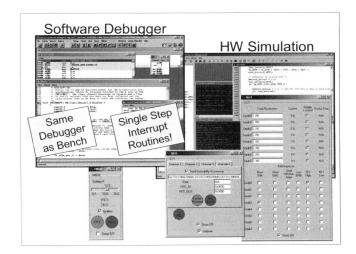

Figure 4 – Screen shot showing execution of the virtual system

The CPU model (also known as the Virtual Processor Model - VPM) executes the target instruction stream cycle-by-cycle, such that register and memory accesses occur just as in the actual hardware. An example of this execution is shown in Figure 4, where the virtual system simulates while the software and hardware are monitored by the debugger. The CPU model by itself executes extremely fast at nearly 100 million instructions per second (on a 2.2GHz Windows 2000 PC). Execution of the CPU and the peripherals slows the simulation to around 10 million instructions per second, which is within one order of magnitude the speed of the actual hardware!

The various ASIC devices range in complexity from a simple serial EEPROM to a deployment ASIC with a dozen control/status registers and an integrated state machine. The most complex device is described with a 60 page functional level specification. Almost all of the devices communicate to the microcontroller via an 8 or 16-bit SPI (Serial Peripheral Interface) synchronous communication. Behavioral level 'C' is used to interconnect the ASIC devices because they use analog voltages not easily represented by Verilog interconnections. Moreover, the ASIC models are connected to the microcontroller model via regular discrete time intervals and by "microcontroller pin events". The "pin events" can occur at any time within the discrete time intervals, and ASIC models process the events in the order they occur. Typically the events are discrete I/O such as chip selects, SPI transfers, and other control signal changes.

For the air-bag system the ASIC discrete time interval was chosen to be the same as the 10kHz sampled accelerometer data, (ie. 100us). The accelerometer model thus supplies a new data sample to the A/D pin of the microcontroller at each discrete time interval. Other ASICs update filtered analog voltages or execute timed events based upon the 100us discrete interval. The choice for the discrete interval can be modified by the end user but should remain less than the fundamental loop time (hundreds of us) of the target software.

The graphical user interface executes at a slower fixed interval. The exchange of data with the graphical user interface is limited to a 1ms interval time to reduce the communication overhead, yet provide for adequate response time of the display.

VIRTUAL SPI EXAMPLE – As described previously the models are written at a cycle-approximate level. An example of this abstraction is the combining of serial communication (bits) into a parallel transfer. The air-bag virtual system makes extensive use of a "SPI streams interface". The high level SPI protocols (such as phase and polarity) are checked but without the implementation of a bit-by-bit electrical interface as shown in Figure 5.

The SPI data is transferred with parallel transfers. The MOSI (Master Out Slave In) and MISO (Master In Slave Out) signals transfer the data bits in parallel. CS (Chip Select) is a normal discrete signal. The CONTROL signal contains several bit fields that allow error checking of the transfer including size, strobe, bit order, clock polarity and phase. This communication of the SPI protocol between the master and the slave devices allows the simulation to warn for miss-matches in the communication protocol. Note that the "Electrical" parameters such as setup and hold time are not communicated nor checked.

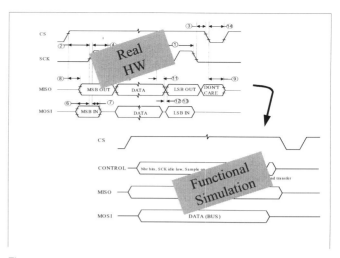

Figure 5 – Virtual SPI "streams interface"

BENEFITS

The air-bag virtual system offers several benefits for both hardware and software verification. Because the verification is conducted on a model (that can be easily changed), the feedback loop to improve the design is shortened. The following benefits have been realized for the air-bag virtual system:

- System level ASIC verification
- Software verification on a virtual system
- Sophisticated analysis tools
- Internal hardware visibility (observe-ability)
- Advanced test benches

This section will discuss each of these benefits in greater detail.

SYSTEM LEVEL ASIC VERIFICATION – ASIC verification at the system level has a high benefit compared to the modeling costs. Existing models may be combined to create entirely new designs. For example, while the original air-bag virtual system was developed to support software development, a relatively low incremental investment in ASIC models allowed the creation of new virtual systems. The new virtual system is then programmed with target (product) software to validate the requirements of the new ASIC prior to the mask set release. Within the environment the SW engineer has the ability to optimize the functionality of the proposed ASIC along with the software driver prior to ASIC design and fabrication.

SW VERIFICATION ON A VIRTUAL SYSTEM – Software verification on a virtual system is very similar to verification on a physical prototype. The simulator and the virtual processor model (VPM) integrate the same debugger used on the bench. The debugger is provided with full run-time control of the simulation as well the ability to set breakpoints, step through source level code, inspect memory and observe behavior of the microcontroller registers.

In simulation, the entire virtual system (including the test stimuli) is paused by the run-time control of the debugger. This allows the software developer to step into and out of interrupt routines while the software remains in synchronization with the test stimuli. This type of increased visibility is one of the reasons the software engineers often prefer the virtual bench for daily software development [6].

Within a system, "plant models" are the stimuli needed to test the ECU. When testing physical hardware, these plant models often consist of graphical interfaces and scripts that control hardware boxes interfaced to the product. The virtual system needs the same types of user interfaces; however, the hardware boxes can be eliminated because the plant models connect to the virtual system directly. The savings associated with virtual test benches (vs. hardware box driven test approaches) may be on the order of $100K per development station. It is, however, clearly understood that there will continue to be a need for a mix of "real benches" and "virtual systems" for product development. This is necessary to validate the actual physical products and to confirm that the virtual system matches the behavior of the physical product.

For a "physical test bench", emulators are typically used to provide the visibility inside the microcontroller. These emulators often require special silicon "bond-out" chips or Field Programmable Gate Array (FPGA) devices in order to make the internal registers visible to the software developer. As microcontrollers increase in complexity and frequency, extensive engineering is required to accurately replicate the microcontroller. In some cases the development of the emulator becomes technically un-feasible and the software engineer is left with only the "background debug mode" access to the silicon and registers or reduce functionality (or frequency) of the emulator. Verification on a virtual system allows visibility of the simulated registers and removes the difficulty of silicon bond-out chips and high frequencies.

ANALYSIS TOOLS – Software and hardware analysis tools can be applied to the simulation to measure system metrics. These metrics include: interrupt latency, stack depth, code coverage, data traces, bus or serial communication traffic. When ASICs often communicate via the serial peripheral interface (SPI), confirming communication messages on the bench requires use of a four-channel oscilloscope and manual observation of each message on a bit-by-bit basis. Simulations, however, are easily instrumented with a "SPI watcher" module that records the SPI transfers while a chip select is active. At the trailing edge of a chip select, the buffer is output with a simple print statement resulting in a display such as:

```
uP SPI to EE: 0x03 0x00 0x3C 0x00 0x00
EE SPI to uP: Hi-Z Hi-Z Hi-Z 0x9F 0xFF
```

A powerful non-intrusive extension to the simulation streams out details of the instruction path, register, memory and cache usage. A Windows DLL (Dynamically Linked Library) is added to the simulation to direct the tracing and profiling. This extension allows the capture of instruction traces, measure code execution time, measure interrupt latency, and monitor code coverage. Subsequent functions can then receive and process the output [2]. An example display is shown in Figure 6.

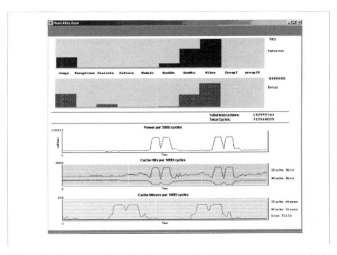

Figure 6 – Analysis of a virtual design using METrix; measuring design parameters such as interrupt service routine latency, cache hits/misses, bus transactions, and code profiling. Data can be filtered to graphics and files and triggers based on either hardware or software events

A minimally intrusive and simple instrumentation method allows a direct connection between the target and the host computer running the simulation. For example, the function TspiHostPrintString() can be added to the target code and provides print capabilities to the host (PC) software window.

One example of this instrumentation is a method to verify the target code implementation. The deployment algorithm is first developed on (and compiled for) a host computer with access to a large library of vehicle acceleration profiles. The host simulation records the algorithm variables every loop time while the acceleration profile is applied, similar to a strip chart. The record from the host computer is then compared to a record generated by the source code compiled for the target system. The record is simple to capture on the host system, but with physical bench equipment an extensive logic analyzer setup is required to capture the many target system variables. Because of the difficult setup, only a limited number of vehicle acceleration profiles may be captured on the bench. Using the direct connection between the target simulation and the host computer the target system can access the library of acceleration profiles directly facilitating a one-step regression process with the large library.

VISIBILITY OF THE HARDWARE – Another benefit of a virtual system is internal hardware visibility. The microcontroller and ASIC simulation models are built with extensive warnings that monitor the execution of the software. For example, the internal EEPROM peripheral model knows when it is in the "write" state and thus a subsequent "read" of the same EEPROM returns invalid data. Many internal EEPROMs are designed to supply high voltages to the internal memory cells during write sequences, and thus data reads are not possible during the several millisecond "write" state. The physical part offers no indication of this illegal (read during a write) condition, and an incorrectly written system may compute critical algorithms using invalid data. The model of the EEPROM is constructed to output a message under this invalid condition:

```
"WARNING: EEPROM1 data read from
address 0xXXXX while in the write
state".
```

This message alerts the software engineer of the specification violation.

An ASIC model may also embed warning messages. Take for example an output driver chip that provides for a current measurement, but while the ASIC is in the measurement state, the die temperature rise is increased. The design of the ASIC package allows for short duration current measurements on a single output driver. The model for that ASIC is built in a way that if the duration is exceeded or the number of measured channels is exceeded, a warning message is displayed to the software developer. In this manner, the assumptions about the usage of the chip are communicated and enforced throughout the development process. The software developer thus benefits from a "field application engineer in the box" that is constantly checking the software execution for correct interaction with the hardware design.

The number of warnings provided in a peripheral or ASIC simulation varies with the complexity of the model. Typically 2 to 20 warning messages are included in each microcontroller peripheral or ASIC model. Many of these warnings monitor subtle requirements and caution statements in the specification. Without these warnings, the target software drivers may be written in violation of

the requirements, which may later result in a quality or field return issue.

Each ASIC model is created with "elements" or registers that provide visibility of the inner functionality. An element may be read only (allowing visibility) or read/write (allowing visibility and control). For example, an accelerometer model provides a writable element of acceleration to modify the force in units of gravity (G's). A power supply model, as shown in Figure 7, can display the contents of internal registers as well as visual indicators of the outputs of the ASIC.

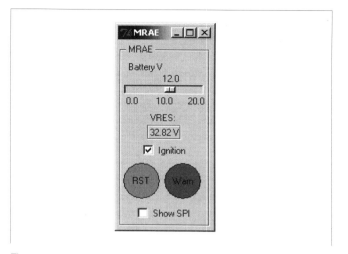

Figure 7 – Example of power supply ASIC showing control of the battery voltage and status of the charge pump (VRES), reset output and warning lamp output.

ADVANCED TEST BENCHES – The virtual system test bench provides the ability to view and control the simulation. A test bench can be as simple as a command line, a script interpreter, or even a full graphical interface. In Figure 3, the graphical interface for the air-bag system is shown on the left. The plant models include acceleration data files; external serial-communication based sensor models, as well as CAN communications. Because a virtual system can be tested without building hardware interface "boxes" it costs significantly less to instrument the virtual system than the actual product. In general, the visual and scripting capabilities of the test bench are shared with the actual bench, i.e. they use the same graphical and scripting environments. Automatic tests thus developed for the virtual test bench also execute without significant modification on a physical test bench.

Automatic regression and unit testing is the ability to regress the functional units of the system at every step of development, looking for unexpected changes introduced by recent modifications. The plant models, debugger, and the simulation share the same host computer system, which facilitates unit testing. Serial communication messages can be generated and intercepted via scripts. The software development team is thus provided with the tools to allow daily execution of the regression test suite. The tests also include the environmental stimuli needed to exercise the software and report a pass/fail upon completion of each test.

Another powerful capability of the test bench is the simulation of multiple models (or actual product) in parallel, with the same test stimuli and similar monitoring functions. For example a high level functional model ("golden model") of the system can by executed in synchronization with a detailed cycle-accurate model, as shown in Figure 8. The test bench can verify parameters of interest such as ignition response times, communication messages or air-bag deployment times by comparing the two models. Although the two models run at very different abstraction levels, their top-level functionality will match to verify implementation. One example is an OEM supplied "golden" model in Matlab, compared with the cycle-accurate target code on the microcontroller and hardware models.

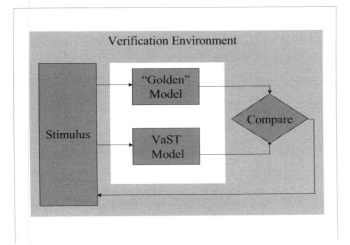

Figure 8 – Golden model comparison

Simulation of the hardware and software improves the quality of the overall system. The "field application engineer in the box" provides continuous enforcement of the proper usage of the hardware. The analysis tools allow visibility across the software/hardware boundary and generation of charts for throughput and code coverage. Elements provide for visibility inside the ASICs, deepening the hardware understanding. Finally the test benches facilitate automatic regression and unit testing.

LOGISTICS

Various logistical aspects must be managed in order to make effective use of product simulation. While it is beyond the scope of this paper to address rollout of these tools, it is important for the reader to see that tool availability, training, and model attainment must be considered.

Virtual systems are easily distributed to worldwide development sites. When intense bursts of software development resources need to be added, it is only a matter of minutes to make another copy of the development environment. A global floating license pool of simulators and models allows virtual system to be invoked worldwide and around the clock. Compilers/linkers and debuggers are the same as those used for the bench environment and can also benefit from widespread availability. Contrast the virtual system availability with the conventional development environment; it often takes a week or more to add a new development "bench".

The virtual system is easy to setup and use, and requires minimal training. We rely on an initial 2-hour individualized training/project start. In this training the basics of invocation and customization of the simulation are covered. That training is supplemented with 1-hour ad-hoc "classes" on detailed topics as well as on-line reference pages. For software engineers already familiar with the debugger, only minimal training is required to become proficient in the use of the simulator.

Clearly, the creation of a virtual system is a large undertaking and requires careful planning. Once the initial virtual system has been created, it will serve as the springboard for future derivative systems. The virtual system consists of microcontroller "core" (VPM and memory devices) microcontroller peripheral models (typically from the semiconductor supplier), ASIC models (typically developed internally) and plant models (typically re-using existing bench test apparatus).

The microcontroller VPM is verified against semiconductor vendor supplied test vectors to assure accuracy. The VPM can be built and verified in the duration of 6-20 weeks, depending on the complexity of the CPU. Microcontroller peripheral models such as the timers, SPI, and A/D are written in behavior level ANSI-C. The peripherals are interconnected to the VPM via one or more "virtual buses". The semiconductor vendor is in the best position to create and verify the peripheral models, as they have the best understanding of the specifications and intricate details

The system ASIC devices often contain company intellectual property. A core team familiar with the model generation tools and the abstraction level required creates the models from the functional specifications. For new devices, the design, source code for the model, and test cases may be shared with the semiconductor supplier as an executable reference. The intent is that the executable reference provides a "golden" model of the ASIC that has been tested in the system implementation. We are working with the various semiconductor vendors to integrate the executable reference model into their respective design flows. For reference, the average ASIC device of the air-bag system required approximately 6 man-weeks for the design, code and verification activities.

RESULTS

Embedded ECU's continue to increase in both software and hardware complexity. Even as transistor densities increase, the bulk of the system complexity will be added to software. In other words, if one compares the complexity of a 20K ECU to a 2M ECU, the complexity of the 2M software is 2 orders of magnitude larger than the 20K ECU, while the complexity of the hardware (CPU, peripherals and ASIC) may seem to be only 1 order of magnitude more complex [7]. A virtual system, once developed, can easily grow to match the memory size changes of the system. Moreover, we have shown the benefits of verifying software on a virtual system as opposed to a hardware platform in terms of time, effort and scalability.

Overall, the entire simulation of the air-bag system with the core, peripherals, ASIC models and test bench executes at a rate of about 1 million instructions per second, approximately 1/15th the speed of the actual hardware. This speed is significantly faster than what is achievable by the typical instruction set simulators and Verilog simulations. The speed is satisfactory for most software development, but may still be slower than desired for 24/7 repetitive regression testing of a system.

The following benefits have been realized during the air-bag product development, due to the use of the virtual system:
- Quickly gathered throughput data for a business decision on new product directions.
- Six hardware schematic issues were found and eliminated prior to board fabrication. Re-work was avoided.
- 27+ target SW issues were found, many of which may have been difficult to detect on the bench. Some examples:
  1. SPI data size changed w/o re-initializing SPI HW
  2. SPI TXE/RXE were modified while PWR=1 (specification violation)
  3. Inter-message communication gap (5ms) to ASIC violated
  4. CAN error interrupt handler did not handle all sources of interrupt
  5. Improper I/O port initialization order
  6. Improper A/D frequency (prescaler) selected causing incomplete conversion result
- Reduce the number of hardware test benches needed to verify the system.

- Demonstrated capability to reduce the design cycle time.

FUTURE VIRTUAL SYSTEMS

Once a baseline system has been modeled, it can then be quickly modified for future virtual systems. An example modification is given as a case study below.

The engineering design group planned for a next generation ASIC that combined two previous ASICs (See Figure 9). The new ASIC removed some diagnostics functionality such that diagnostics previously done in hardware are now implemented via software algorithms. This change impacts the target software because the diagnostics must be manually commanded and read via the microcontroller. While paper studies were conducted on this new approach, some lingering questions remained about the SPI traffic loading and the impact of the software diagnostics on the microcontroller throughput. Those questions were easily answered through the analysis of the application software on the virtual system.

The new ASIC specification was provided to the semiconductor supplier in February 2004. The hardware development schedule is about 10 months to create the actual ASIC, and 2-3 additional months to build the hardware prototype, write the software, and measure the impact.

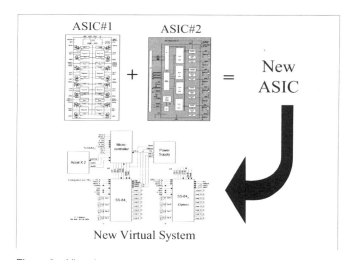

Figure 9 – Virtual system can measure a planned system change with significantly less expense than the development of hardware

In parallel with the ASIC design activity, we modeled the new ASIC and integrated it within the existing virtual air-bag system. Instead of the 10 months required for the silicon design, the modeling activity was completed in about 10 weeks. That is about the same time required to write and review the new software requirements for the device. Then, using the virtual system, the actual software drivers were implemented. The use of the virtual system during this development resulted in 8 target code design and coding error corrections and allowed for a 1-week integration/testing time of the silicon with the (already verified) target software.

The process of creating the ASIC model identified 9 functional changes that were implemented in the first pass silicon along with more than 30 specification corrections/ clarifications. Clearly, the addition of up-front modeling adds a rigor that improves the quality of the design as well as the time to market.

CONCLUSION

In this paper we have presented a virtual system ECU that encourages development and testing in an environment that significantly contributes to the quality. The virtual system executes the same code as used on the hardware bench, at approximately $1/15^{th}$ the speed of the real system. Full functionality is supported, including the ASICs and various plant models to stimulate the system. The built-in "field application engineer in the box" warnings help the software engineer to identify various required target software changes.

The realized development time savings are associated with the elimination of design rework, such as:

- Evaluation of new ASICs prior to the expense of silicon development
- Development of driver software in parallel with silicon and board fabrication
- Detection of difficult to find software driver defects via: "field application engineer in the box"
- Comparison of a golden reference to a nearly fully implemented system
- Regression and unit testing as the ECU is being built rather than after it is complete
- Visibility into details of the hardware operation
- Analysis tools for system throughput, interrupt latency, code coverage and communication traffic
- Savings of bench costs through replacement of the pre-silicon emulator and equipment costs

In summary, the virtual system allows the software and hardware activities to be parallelized, resulting in less total manpower, and reduced time-to-market development cycle for the creation of a new ECU.

ACKNOWLEDGMENTS

We gratefully acknowledge the end users of the air-bag system simulation who provided near daily feedback and improved usability of the simulation.

REFERENCES

1. Cadence Virtual Component Co-Design (VCC) environment for engine control unit design: http://www.cadence.com/company/success_stories/magneti_ss.pdf
2. James Brogan "Guide to Creating a User METrix DLL for CoMET and METeor Toolsets", 22-Apr-2003, VaST Systems Technology.
3. Lennard, Chris; Mista Davorin; "Taking Design to the System Level", April 2005, http://www.arm.com/pdfs/ARM_ESL_final_checked(3)JC.pdf
4. "Maximizing Silicon ROI: The Cost of Failure and Success", Nassda White Paper WP020522-1A; http://www.nassda.com/ROI_WP_v062503.pdf
5. "The verities of verification", Bill Roberts, Electronic Business 01-Jun-2003; http://www.reed-electronics.com/eb-mag/article/CA301056?industryid=2112
6. Perrier, Vincent; "A look inside electronic system level (ESL) design", 26-Mar-2004, EETimes, http://www.eetimes.com/news/design/features/showArticle.jhtml?articleID=18402916
7. Winters, Frank; Mielenz, Carsten; Hellestrand, Graham, "Design Process Changes Enabling Rapid Development", Convergence 2004, http://www.vastsystems.com/notes/convergence20041018.pdf

CONTACTS

Everett R. Lumpkin (Simulation and Modeling) 765-451-3247; everett.r.lumpkin@delphi.com

Michael T. Gabrick (Digital IC design) 765-451-3056; michael.t.gabrick@delphi.com

DEFINITIONS, ACRONYMS, ABBREVIATIONS

ASIC – Application Specific Integrated Circuit

Co-Design – The process of exploring various CPU architectures and system partitioning. Also referred to as the simultaneous design of hardware and software.

Co-Verification – The process of verifying the correct operation of the system, including hardware, software and the timing effects of each, typically at the cycle-accurate or cycle-approximate level.

CPU – Central Processing Unit, usually used in the context of general microcontroller architecture without a specific embedded peripheral set.

Cross-compiled code – Use of a host computer to compile, link and locate code for a target computer.

Cycle-Accurate Simulation – Behavioral model of system hardware (CPU & peripherals) with timing steps equal to the fundamental bus frequency.

Cycle-Approximate Simulation – Behavioral model of the system hardware (CPU & peripherals) with timing that approximates the behavior of pipelining and cache effects.

ECU – Embedded Control Unit, sometimes also referred to as an Electronic Control Unit.

ESL – Electronic System Level design

Executable Specification – A host platform functional model of the system that allows customer and test interaction with the product.

GUI – Graphical User Interface

Golden Model – A reference model used to convey the functionality of the system. Used to compare with a more detailed model.

HDL – Hardware Description Language, i.e. Verilog or VHDL.

HLL – High Level Language, i.e. C/C++.

Host – Engineering workstation commonly using HPUX, Solaris or Windows operating system.

ISS – Instruction Set Simulator, a model of a CPU that is able to execute cross-compiled application software.

Module – Grouping of related functions to perform a requirement of the system. Also known as a block and may become either hardware or software.

NRE – Non-Recurring Engineering Costs

Plant – A portion of the model that simulates external stimuli such as sensors, actuators, or even a vehicle transmission.

Regression Test – A suite of tests executed against the product that can automatically determine if the product has "regressed" from a previous level of functionality due to additional detail/refinement added to the model. Sometimes known as test vectors.

RTL – Register Transfer Level, a level of abstraction of digital hardware designs that allows synthesis from HDL code to a gate level representation.

Target – Embedded CPU such as ARM, PowerPC, Star12, V850, or TriCore.

Test Bench – Simulation environment that surrounds the system model to supply input stimuli and displays behavior of the system. May support command line or graphical user interfaces.

Unit Test – A detailed test of a subsystem. Functional unit tests determine if a module is within the capabilities of the design. Structural unit tests (white box) determine of a module is correctly coded.

Verilog – A hardware description language

VHDL – VHSIC (Very High-Speed IC) Hardware Description Language

VPM – Virtual Processor Model

Using VHDL-AMS as a Unifying Technology for HW/SW Co-verification of Embedded Mechatronic Systems

Thomas R. Egel
Mentor Graphics

Norman J. Elias
Consultant

ABSTRACT

The low cost of microcontrollers makes them increasingly popular for electronic control of a wide range of embedded systems throughout the vehicle. An embedded mixed-signal or mechatronic system is one that uses a microcontroller to control some aspect of the physical system such as motion, speed, temperature etc... The added dimension of software presents a significant challenge to the design, integration and verification of this class of systems.

This paper will discuss how the VHDL-AMS language can be used to describe the behavior of the physical hardware along with the controlling software algorithm together in a unified system model. This unifying technology can provide invaluable insight to the embedded system designer during the system design, integration, verification and debugging phases.

INTRODUCTION

Created in response to industry demand for a non-proprietary modeling language and approved by the IEEE in 1999, VHDL-AMS is the only industry standard, open-source hardware description language[1] capable of modeling diverse systems consisting of both electrical and non-electrical components. Being a superset of the all-digital VHDL language also makes it highly capable for controller hardware and software modeling. The analog and mixed-signal extensions (AMS) provide the enabling technology that allow the engineer to take the "next step" in HW/SW co-verification with the inclusion of both electrical and non-electrical hardware in the system model.

As system requirements become more complex and demanding, the use of microcontrollers continues to increase. Current system integration methods, which rely on the availability of the actual hardware, occur very late in the design cycle. Any problems that arise at this time often result in expensive design or schedule changes. Having a VHDL-AMS system model that incorporates both the hardware and software can allow a "virtual" system integration phase to begin well before the actual hardware is available. Such a model can help the system designer detect unforeseen integration issues and allow alternatives to be explored much earlier in the design cycle.

EMBEDDED MECHATRONIC SYSTEM DESIGN CHALLENGES

While the core of any embedded system is the microcontroller, the design of embedded mechatronic systems is further complicated by the inclusion of heterogeneous technologies. Figure 1 represents a typical embedded mechatronic system consisting of sensors to convert a physical quantity (such as position, temperature or speed) to an electronic signal for digital processing by the microcontroller.

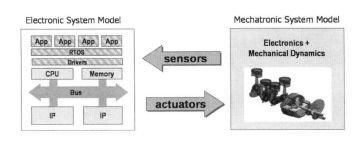

Figure 1 - Embedded Mechatronic System

Actuators receive the electronic signal from the controller and produce a physical output (such as mechanical motion). A powertrain control system, for instance, is comprised of hydraulic valves and actuators, moving mechanical gear sets, all precisely controlled electronically by an embedded microcontroller. A cruise control system processes information from sensors to determine the vehicle speed and adjust the throttle as needed. Even a "simple" voltage regulation system can use sophisticated current and temperature sensing techniques for precise voltage control.

VHDL-AMS is a powerful language capable of modeling all aspects of these embedded mechatronic systems. Design engineers can use the language to create detailed math-based hardware models of the individual components. System engineers can combine these models to perform "virtual" system integration. The following section describes in more detail how VHDL-AMS can be applied to solving the unique problems that embedded mechtronic systems present.

HARDWARE MODELING WITH VHDL-AMS

To be useful for system design, integration and verification, a modeling language must provide constructs that facilitate both top-down and bottom-up design methods. For system simulation to be effective, detailed device models must co-exist with high-level abstract models. In addition, a mechanism must exist allowing the user to easily select the appropriate level of detail for the analysis task at hand.

VHDL-AMS provides the ability to describe a system component at the appropriate level of abstraction allowing optimal model detail to be included at any stage of a top-down or bottom-up design process. In the early stages of a top-down design process, high-level models can be used to verify the system performance will meet the desired specifications. Here modeling the details of the technical interactions may be postponed until further definition becomes available. As the design progresses, technology-specific models can be inserted as they are defined. Towards the end of the design cycle, bottom-up verification techniques can be used to verify the entire system complete with multiple technologies. The result is the ability to perform successive verification in an orderly fashion as the design definition progresses.

Unlike other system modeling methods, VHDL-AMS is capable of both signal flow and conservative modeling. Signal flow or transfer function modeling techniques are commonly used in high-level block diagrams. Individual blocks have inputs and outputs and the data "flow" is unidirectional. While useful for verifying system behavior at a high level, this approach is extremely limited for modeling physical hardware interactions. Conservative modeling techniques, however, apply the laws of conservation at the connection points between models. Here data flow is bi-directional, automatically accounting for hardware interactions, such as loading effects. As a result, the conservative approach is more "hardware oriented" in that models can be developed to behave like the actual hardware without any special modeling considerations. Conservative models can also be easily created to accommodate parameters that directly correlate to the physical hardware. The ability to combine both modeling techniques within a single system model is one of the strengths of the VHDL-AMS language. In addition, the multi-technology constructs in VHDL-AMS provide the ability to model analog and

digital electronics interacting with non-electrical components in the context of a complete system. As such, it is uniquely positioned to address the needs of engineers designing today's complex mechatronic systems.

The structure of a VHDL-AMS model[2] consists of an *entity* and one or more *architectures*. The entity defines the physical connection points along with any parameters needed for characterization. The architectures describe how the device operates using equations, algorithms or even extracted device data.

The sections that follow will provide some example models to illustrate how the VHDL-AMS language provides a unifying technology with the power and flexibility needed to solve complex system design and analysis tasks.

HIGH-LEVEL MODELING

In the early stages of a top-down design process, high-level modeling techniques can provide useful insight into the overall system performance. VHDL-AMS provides constructs to create transfer function models in both the s-domain and z-domain. An example of a simple lead-lag block modeled in the s-domain is shown in Figure 2.

This model has an input, output and parameters for the gain, pole and zero frequencies. It would typically be

```
entity q_LeadLag is
  generic (
    K   :   real := 1.0;      -- gain
    Fp  :   real := 20.0e3;   -- pole frequency (in Hz)
    Fz  :   real := 1.0e6;    -- zero frequency (in Hz)
  );
  port (
    quantity input  : in  real;
    quantity output : out real);
end entity q_LeadLag;

architecture s_dmn of q_LeadLag is
  constant wp  : real := math_2_pi*Fp; -- Pole freq (in radians)
  constant wz  : real := math_2_pi*Fz; -- Zero freq (in radians)
  constant num : real_vector := (1.0, 1.0/wz); -- Numerator
  constant den : real_vector := (1.0, 1.0/wp); -- Denominator
begin
  output == K * input'ltf(num, den);   -- Laplace transform of input
end architecture s_dmn;
```

Figure 2 - VHDL-AMS Lead-Lag Model

used with other signal flow models in a block diagram system model during the early stages of design definition as shown in Figure 3.

The power of VHDL-AMS is that these high-level blocks can eventually be replaced by physical hardware models and verified within the context of a complete system. The advantage of this will be clear as our discussion continues.

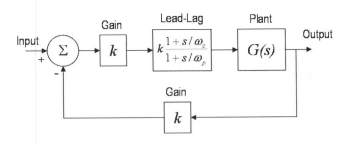

Figure 3 – Block Diagram System Model

ANALOG, DIGITAL AND MIXED-SIGNAL MODELING

In a typical mechatronic embedded system, the electronic hardware is often a combination of analog and digital circuitry, each of which requires significantly different modeling techniques. The electro-mechanical plant is primarily analog but the controls are often digital. Unlike continuous analog quantities, digital signals assume discrete values as determined by the control algorithm. Digital signal transitions constitute events that occur in a time sequence. This concept of event-driven digital signals is an essential element of a system modeling language. VHDL is a well-established digital hardware description language fully capable of modeling all aspects of digital electronics. The AMS extensions incorporate the necessary analog constructs making VHDL-AMS the unifying "glue" that links these distinctly different domains. Figure 4 shows examples of analog and digital waveforms created from a VHDL-AMS simulation.

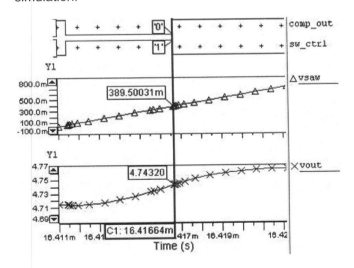

Figure 4 - Analog and Digital Signals

The analog waveforms (*vout* and *vsaw*) are continuous in the Y-axis and are calculated by the simulator using variable time steps (as indicated by the symbols). The digital signals (*sw_ctrl* and *comp_out*) have discrete Y values (0 and 1), which are assigned by the simulator when events occur. You can see that the analog "activity" increases during digital transitions. The VHDL-AMS language includes constructs to ensure that the analog and digital signals stay synchronized during a

simulation. VHDL-AMS is unique in that analog and digital behavior can be modeled and synchronized within a single industry standard language. As we will see, this is a crucial requirement when modeling embedded mechatronic systems.

MIXED-TECHNOLOGY MODELING

To fully understand an embedded mechatronic system, the non-electrical components like sensors and actuators must be taken into account. Sensors convert some physical quantity (i.e. speed or temperature) into an electrical signal for processing. Actuators convert an electrical command signal from the processor into some physical phenomena (i.e. motion). Only with accurate models of these non-electrical devices interacting with the electronics entirely in a closed loop system can proper analysis be achieved. Using VHDL-AMS, system integration can begin before actual hardware is available by creating a system model that incorporates a combination of the various technologies. They may include the mechanical, magnetic, hydraulic, thermal effects or any other technology that can be described using algebraic differential equations.

VHDL-AMS has distinct advantages over traditional system modeling methods for modeling hardware of various technologies. To illustrate this let's examine the model development process for a simple DC motor. The starting point for any model is the equations that describe the behavior of the component or system. A DC motor is a mixed-technology (electromechanical device) governed by the following device equations[4]:

$$T = -K_t \cdot i_a + D \cdot \omega_m + J \cdot \frac{d\omega_m}{dt} \qquad (1)$$

$$V = K_e \cdot \omega_m + i_a \cdot R + L \cdot \frac{di_a}{dt} \qquad (2)$$

Where T is the motor torque at the shaft and V is the voltage across the electrical terminals.

The typical block diagram modeling approach would be to develop a transfer function model. This involves taking the Laplace transform of these equations and solving for the appropriate variables. After much manipulation, the following equation relating the voltage (V) torque (T) and angular velocity (Ω) is derived:

$$\Omega_m(s) = \frac{A}{\tau \cdot s + 1} V_a(s) + \frac{B}{\tau \cdot s + 1} T_q(s) \qquad (3)$$

Where:

$$\tau = \frac{J(R + sL)}{D(R + sL) + K_e K_t} \qquad \textit{(Time Constant, sec)}$$

$$A = \frac{Kt}{D(R+sL)+K_eK_t} \quad \text{(Voltage Gain, rad/volt-sec)}$$

$$B = \frac{(R+sL)}{D(R+sL)+K_eK_t} \quad \text{(Torque Gain, rad/N-m-sec)}$$

Further manipulation is required to represent this in block diagram format, as shown in Figure 5, so it can be included as part of the overall system model

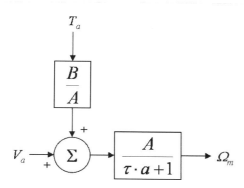

Figure 5 - Block Diagram DC Motor Model

Even for this relatively simple device, the generalized transfer function approach results in a significantly obfuscated model. The original device equations are nowhere to be found and the device parameters are buried within the transfer function coefficients. Furthermore, any change in the original equations will require significant rework to derive the new transfer function.

By contrast, the VHDL-AMS model for the same DC motor is shown in Figure 6. Here the original device equations (1) and (2) are preserved within the architecture of the model and device parameters are clearly identified and passed in through the entity declaration.

```
entity DCMotor is
  generic (
    r_wind : resistance;    -- Motor winding res [Ohm]
    kt   : real;            -- Torque coeff [N*m/Amp]
    ke   : real;            -- Back emf coeff [V/rad/sec]
    l    : inductance;      -- Winding induc [Henrys]
    d    : real;            -- Damp coeff [N*m/(rad/sec)]
    j    : moment_inertia); -- Inertia [kg*meter**2]
  port (terminal p1, p2 : electrical;
        terminal shaft_rotv : rotational_velocity);
end entity DCMotor;

architecture basic of DCMotor is
  quantity v across i through p1 to p2;
  quantity w across torq through shaft_rotv to ref;
begin
  torq == -1.0*kt*i + d*w + j*w'dot; -- Torque Equation
  v    == ke*w + i*r_wind + l*i'dot; -- Voltage Equation
end architecture basic;
```

Figure 6 - VHDL-AMS DC Motor Model

Any changes in the original device equations or parameters can be directly implemented in the model with minimal effort. In addition, this model can be shared among other team members with little or no additional documentation.

MODELING ABSTRACTIONS

Another key requirement for effective system modeling is to have the ability to describe the components at various levels of abstraction. This allows the engineer to focus on the details of a specific part of the system, while maintaining its context within the overall system design. This is particularly useful for top-down design methods, where the detail is added as the system design progresses. During the system definition stage, block diagram or signal flow methods are often used to describe the major aspects of the system. These methods are effective for system partitioning, but fail to completely account for interactions among the blocks. VHDL-AMS provides the ability to use block diagram methods for high-level modeling while selectively substituting physical hardware models as desired. Additionally, higher-level models can be incorporated when using VHDL-AMS for bottom-up verification to optimize the simulation speed and accuracy. Also known as "checkerboard verification", this is only possible with a modeling language capable of both signal flow and conservative modeling. In addition, VHDL-AMS provides this capability using the concept of multiple architectures which can be selected at simulation run time.

As an example, when designing a speed or position control system, the VHDL-AMS DC motor model from the previous section can easily be inserted into the control loop as shown in Figure 7.

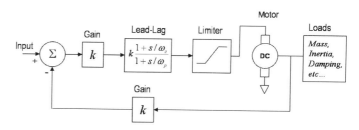

Figure 7 - Servo Control Loop with DC Motor

As the other blocks become more defined, they can be replaced by more detailed hardware models. During this verification process, realistic loads can be attached to the motor to examine the effects of loop stability. VHDL-AMS provides the ability to create models of varying degrees of fidelity and include them at any stage of the design process.

As we have shown, VHDL-AMS provides engineers with a modeling language for all physical aspects of a mechatronic system at arbitrary levels of abstraction. For an embedded mechatronic system, it is also necessary to incorporate the software algorithm into the system simulation to verify the overall system performance. The remainder of this paper explores how VHDL-AMS can be used for hardware/software co-verification of this class of systems.

MODELING THE MICROCONTROLLER

For an *embedded* mechatronic system, "virtual" integration of hardware and software requires a model of the microcontroller executing the software instructions that control the overall system performance. The challenge is to incorporate the software algorithm into the flow of the simulation.

MICROCONTROLLER MODELING ABSTRACTIONS

During system integration the microcontroller can be modeled as an algorithmic block. A single VHDL architecture can be constructed to model the software algorithm and verify its performance in a simulation at the block diagram level.

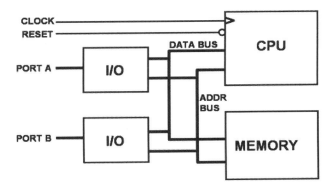

Figure 8 - Microcontroller Core

However, the algorithmic model is inadequate at more advanced design stages. Interactions between the physical hardware and the embedded software can only be properly analyzed if the software timing is accurately modeled. A gate level or RTL model of the microcontroller structure will model software timing to the precision of the system clock but is prohibitively complex and slow. A more realistic approach is to model the microcontroller as a finite–state machine that executes software instructions in sequence producing the correct pin outputs at accurate time intervals. Such a model is referred to as an *Instruction Set Model* [3] of the microcontroller.

INSTRUCTION SET MODELING

As shown in Figure 8, the core of a microcontroller is comprised of a CPU for interpreting and executing software instructions, memory for storing the instructions and data, and I/O controllers for feeding data into and out of the microcontroller.

VHDL architectures of the I/O and memory are straightforward applications of digital modeling. The CPU model, however, is much more complex in concept. Its task is to execute the software instructions which are stored in memory as a sequence of binary numbers.

For each instruction, the specific tasks involved are as shown in Figure 9, namely:

1. Decode the sub-sequence of numbers stored in the memory for that instruction.

2. Locate and read all data values required to execute the instruction.

3. Execute the instruction to determine the results.

4. Synchronize timing to match instruction delays.

5. And finally store those results after all of the above processing is complete.

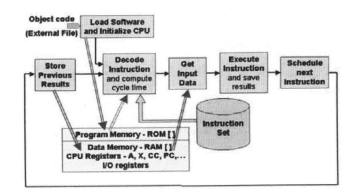

Figure 9 - Sequence of CPU Operations

The physical CPU takes a finite amount of time to complete these operations. The time required varies, depending on the instructions, and is determined by the number of CPU clock cycles required to perform the operations. To accurately simulate system performance the CPU model must account for these delays. The VHDL model can do so by simply counting the required number of clock cycles before storing the results of each instruction. VHDL provides a "wait" statement that is a natural mechanism for implementing this CPU scheduler.

Microcontroller reference manuals published by the manufacturer universally include descriptions of the complete instruction set. The data provided for each instruction is public information and is readily incorporated into the CPU model as a database for interpreting the control software.

PROGRAMMING THE MODEL

A requirement for any instruction set model must be the ability to read the software. In the real device, the software is embedded in the hardware by compiling or assembling the source code and subsequently burning the resulting object code into the microcontroller memory. An accurate simulation requires that the microcontroller model execute that exact same object code. This is accomplished by incorporating an object code file reader into the model which loads the compiled software into the simulated memory. As with the physical system, different versions of the software can be simulated by recompiling and loading the modified code. But, unlike the physical system, the simulation can quickly test numerous software variations without repeatedly burning code into chip samples and thereby consuming stockpiles of chips. Moreover, the "virtual" CPU is safe from the hazard of destructive failures. Figure 10 illustrates the compilation process for both the real microcontroller and simulation model.

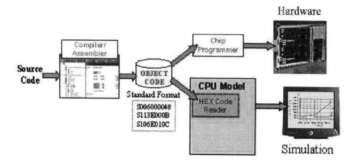

Figure 10 - Compilation Process

As you can see, the same compiled code used by the actual chip programmer can also be read by the VHDL CPU model. This provides extra assurance that when a prototype system is finally built, it will behave as expected due to the analyses performed using the system model.

PROBING THE MODEL

Another advantage of having a VHDL instruction set model of the processor is the ability to probe not only the input and output signals but also the internal software "signals". Figure 11 shows an example of these signals from a simple test circuit created to exercise some of the CPU instructions.

As you can see, an analog output (*out_val*), digital carry signal and the status of the internal accumulator can all be viewed together and values compared over time. Closer examination reveals, the analog ramp increasing in discrete steps but with an apparent negative step at about 86us. To explain this we can look at the microcontroller internal signals. The carry bit seen at the time of the step indicates a numerical overflow, however,

the overflow appears to precede the step by a fraction of a microsecond.

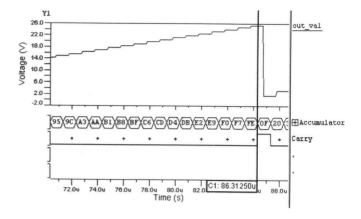

Figure 11 - Microcontroller Signals

Probing deeper still into the actual software operations, illustrates the full power of the instruction set model. The waveforms in Figure 12 include a software "signal" that identifies the specific instructions executed by the microcontroller. This expanded view of the time scale shows that the carry overflow is the result of an *addition* instruction (ADDA) which increments the Accumulator beyond its maximum value. The analog output step does not appear until after this Accumulator signal is propagated to the output port by the *store* (STAA) instruction. The time between the carry overflow and the output step is a delay due to hardware and software.

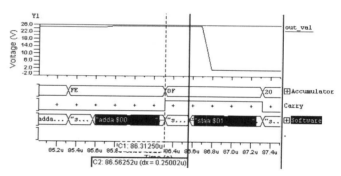

Figure 12 - Hardware and Software Details

Using simulation of the instruction set model for HW/SW co-verification allows us to identify this behavior qualitatively and to analyze it quantitatively. Access to the internal registers of the instruction set model is not possible using a physical processor. This added access can provide invaluable insight into the performance of the processor with the overall system

SUCCESSIVE VERIFICATION METHODOLOGY

One of the challenges in designing embedded mechatronic systems is the difficulty in debugging and verifying the operation of the software within the context of the complete system. Typically, the true system performance is not known until the first physical prototype is built. System integration at this late stage in the design cycle often results in problems due to unanticipated interaction between the hardware and software.

This final section will discuss how VHDL-AMS can be used for successive verification at critical stages of the system design, ranging from specification through final certification and beyond into the introduction and release of design updates. The DC motor servo loop example will be used to illustrate how to apply successive verification techniques using VHDL-AMS.

SERVO POSITION CONTROL LOOP EXAMPLE

Here we will revisit the servo control loop in Figure 7. This control system can represent any number of subsystems within the vehicle where a DC motor is used for position control. Some examples include power mirrors, power seat, and headlamp applications. The individual blocks were initially modeled as transfer functions in the S-domain, but as the design progresses, a decision must be made whether to implement each block in hardware or software.

It has been already shown that VHDL-AMS can sufficiently model any hardware implementation. Here we will explore the additional aspect of including software in the loop. To begin this let's revisit the lead-lag compensator. To implement a function in software, a 4 step design progression can be utilized with VHDL-AMS to perform successive verification as the design evolves:

1. Implement the lead-lag function in the s-domain and verify loop stability.

2. Convert to the z-domain and explore effects of sample clock frequency on loop stability

3. Convert to difference equations and verify system performance

4. Implement difference equations in C-Code and perform final system verification with instruction set microcontroller model.

For step 1, the VHDL-AMS model in Figure 2 was created using the Laplace transform operator ('ltf) to implement equation (4).

$$\frac{Y(s)}{X(s)} = k \cdot \frac{s+a}{s+b} = k \cdot \frac{1+s/\omega_z}{1+s/\omega_p} \qquad (4)$$

Where Y(s) is the output and X(s) is the input. The pole and zero locations are passed into this model as parameters, allowing loop stability to be easily examined as a function of pole/zero locations. For step 2, a bilinear transform is performed by making the substitution for the Laplace variable s as shown in equation (5).

$$s \Rightarrow \frac{2}{T}\left(\frac{1-z^{-1}}{1+z^{-1}}\right) \qquad (5)$$

This is typically a manual process, but can be easily automated within a VHDL-AMS model using the z-domain transform operator ('ztf) as shown in Figure 13.

```
entity q_LeadLag is
  generic (
    K      : real := 1.0;        -- gain
    Fp   : real := 20.0e3;      -- pole frequency (in Hz)
    Fz   : real := 1.0e6;       -- zero frequency (in Hz)
    Fsmp : real := 10.0e3       -- For Z-dmn only: Sample freq (in Hz)
  );
  port (
    quantity input  : in  real;
    quantity output : out real);
end entity q_LeadLag;

architecture z_dmn of q_LeadLag is
  constant T  : real := 1.0/Fsmp;         -- Sample period
  constant wz : real := math_2_pi*Fz;     -- Zero freq (in radians)
  constant wp : real := math_2_pi*Fp;     -- Pole freq (in radians)
  constant n0 : real := wp*wz + 2.0*wp/T; -- z^0 num coeff
  constant n1 : real := wp*wz - 2.0*wp/T; -- z^-1 num coeff
  constant d0 : real := wp*wz + 2.0*wz/T; -- z^0 den coeff
  constant d1 : real := wp*wz - 2.0*wz/T; -- z^-1 den coeff
  constant num : real_vector := (n0, n1); -- numerator
  constant den : real_vector := (d0, d1); -- denominator
begin
  output == K*input'Ztf(num, den, T); -- Z-domain transfer function
end architecture z_dmn;
```

Figure 13 - Z-Domain Lead-Lag VHDL-AMS model

Again this model can be characterized simply by supplying the pole and zero frequencies. The constants n0, n1, d0 and d1 then become the z-domain transfer function coefficients as shown in equation (6). An additional model parameter *Fsmp* represents the sampling frequency which can be varied over a range of values to examine its effects on system performance.

$$\frac{Y(z)}{X(z)} = k \cdot \frac{n_0 \cdot z^0 + n_1 \cdot z^{-1}}{d_0 \cdot z^0 + d_1 \cdot z^{-1}} \qquad (6)$$

To illustrate the advantage of this approach, the system in Figure 7 was simulated at various clock frequencies. The resulting error signal in Figure 14 highlights the effects of the clock on the system error. As the frequency decreases, there is a visible increase in error and loop instability. This analysis determined that a minimum clock frequency of 10kHz is needed to meet the acceptable error criteria.

The 'ztf operator also allows you to perform a final frequency analysis to verify the pole and zero locations will satisfy the system stability criteria. This is important because once the model takes on the form of difference equations, the analysis will be limited to the time domain.

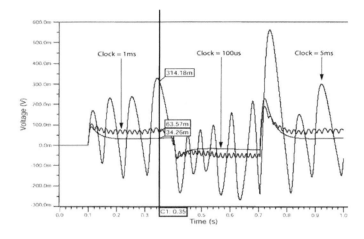

Figure 14 - Effect of Clock Freq on System Error

Step 3 in the process involves converting the z-domain transfer function into to difference equations[2,5]. This is a simple algebraic manipulation resulting in an equation in the form of (7).

$$Y(k) = a_0 X(k) + a_1 X(k-1) + a_2 Y(k-1) \qquad (7)$$

This difference equation can be easily modeled in VHDL-AMS, but more importantly, it can also be directly written in C-code. This leads us to the design goal of implementing the lead-lag block in software using a microcontroller. The successive verification approach outlined here provides the ability to debug the control algorithm as the design progresses. This approach provides a systematic method of ensuring successful system integration well before the prototype stage. Step 4 in this process of successive verification involves implementing the actual C-code and compiling it to a target processor. The combination of an instruction set CPU model and realistic hardware models, provides the engineer with a powerful system model useful for troubleshooting and final verification.

TROUBLESHOOTING

While performance verification is an analysis function, the corresponding design function is troubleshooting. A few of the important troubleshooting tasks that can be performed using simulation are:

1. Analyze destructive and non-destructive failures.

2. Analyze and verify hardware/software interactions.

3. Determine critical timing paths.

4. Guide the optimization of system performance.

The ability to probe internal signals is a powerful aid to failure analysis. Simulation plays an especially critical role in analyzing destructive failures because:

- No hardware is actually damaged.

- The failure is completely repeatable.

- Internal signals can be probed to diagnose the problem.

As we have shown, in using a VHDL-AMS system model to troubleshoot embedded mechatronic systems both the hardware and software details can be probed. Hardware performance can be displayed by viewing signals throughout the electro-mechanical plant and the digital controller. Signals internal to the microcontroller can be displayed to trace the state of the CPU registers. Additionally, the memory buses can be probed to monitor digital data transitions. As a further aid, the simulation model can print a full or partial dump of memory contents at any "time point" in the simulation. It can also print a run-time trace of software instructions to relate hardware events to the progress of the software.

Timing is a critical factor in the integration of embedded software into a hardware environment. The control algorithm has to complete critical functions quickly enough to respond to changes in the physical environment. Excessive software delays can render the system unstable. The capability of the simulator to probe internal signals and trace software execution enables the designer to calculate and analyze control delays. The simulation can distinguish between hardware and software delays and analyze the software delays down to the contribution from each instruction executed. This capability lets the designer expose the critical timing paths and introduce modifications to optimize system performance.

FINAL VERIFICATION

As we have shown, simulation serves a valuable function in verifying system performance at all stages of the design process. This brings us to the final verification step, before a prototype system is assembled for test. The ability to verify the operation of the software running on the CPU within the context of the complete system using successive verification methods, allows the engineer to ensure overall system performance. More importantly, this can be initiated before the hardware is available. As an example, Figure 15 shows a performance aspect of our position control system. The graph compares the final motor shaft position with the input voltage command signal. The system model used here includes the lead-lag filter difference equation implemented in software and compiled to an HC12 microprocessor instruction set model. This model is connected to VHDL-AMS hardware models for the DC motor and mechanical loads. A qualitative (visual) analysis indicates good tracking under static load conditions. Once the nominal operation is verified, additional test benches can be easily created to exercise the model under varying load conditions.

Furthermore, this system model also serves as a vehicle for maintaining design integrity as new features are added to the design. A series of simulation test benches can be defined to test existing features. The simulation commands and parameters can then be saved along with a database of expected results. As new features are added, these simulations can be repeated to serve as a regression test guaranteeing design integrity.

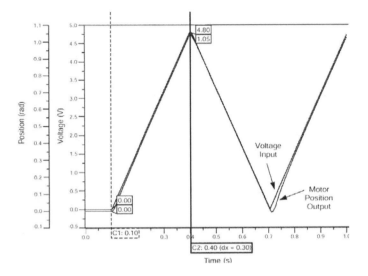

Figure 15 - Voltage Command and Motor Position

CONCLUSION

The combination of software used to control mixed technology hardware in an embedded mechatronic system presents a unique set of challenges for the engineering team. A successful design depends on an understanding of how these technologies interact within the context of the complete system. These interactions are often only fully realized during the system prototype stage, and unfortunately problems uncovered this late can be difficult and costly to correct. The VHDL-AMS modeling language provides a unifying technology for the design and integration of embedded mechatronic systems. The power and flexibility of VHDL-AMS make it superior to existing block diagram methods or circuit simulation techniques for both top-down and bottom-up design processes. This paper has described how a VHDL-AMS system model can be used during all phases of the hardware and software design, from system specification and concept exploration to software debugging, helping to ensure a successful system integration and product launch.

ACKNOWLEDGMENTS

Many thanks to R. Scott Cooper of Mentor Graphics for his guidance and research on this topic.

All simulations were performed using SystemVision™ by Mentor Graphics.

REFERENCES

[1] IEEE-SA Standards Board, "IEEE Standard VHDL Analog and Mixed-Signal Extensions" (Language Reference Manual), IEEE March, 2002.
[2] P.J. Ashenden, G.D. Peterson, D.A.Teegarden, "The System Designer's Guide to VHDL-AMS", Morgan Kaufmann Publishers, 2002.
[3] N.J. Elias, "Instruction Set Modeling of Micro-Controllers for Power Converter Simulation", APEC 2003.
[4] Franklin, G.F., Powell, J.D., Emami-Naeini, A., "Feedback Control of Dynamic Systems", Fourth Edition, Prentice Hall, 2002.
[5] Cooper, R. S., "The Designer's Guide to Analog and Mixed-Signal Modeling", Avant! Corp, 2001.

CONTACTS

Tom Egel is a Technical Marketing Engineer at Mentor Graphics specializing in automotive application development using VHDL-AMS. Previously employed by Analogy, Inc., Tom has over 15 years of experience with various hardware description languages and simulation methods. For more information on this topic, he can be contacted at tom_egel@mentor.com.

Norm Elias is an independent consultant specializing in simulation and modeling of automotive and other mixed-signal, mixed-technology systems. Over a 30 year career including research positions at Bell Laboratories and Philips Electronics, Norm has pioneered simulation technologies for statistical design and for modeling embedded software. For more information on these topics, he can be contacted at norm.elias@ieee.org.

2005-01-1342

Virtual Prototypes as Part of the Design Flow of Highly Complex ECUs

Joachim Krech
ARM

Albrecht Mayer and Gerlinde Raab
Infineon Technologies AG

ABSTRACT

Automotive powertrain and safety systems under design today are highly complex, incorporating more than one CPU core, running with more than 100 MHz and consisting of several 10 million transistors. Software complexity increases similarly making new methodologies and tools mandatory to manage the overall system. The use of accurate virtual prototypes improves the quality of systems with respect to system architecture design and software development. This approach is demonstrated with the example of the PCP/GPTA subsystem for Infineon's AUDO-NG powertrain controllers.

INTRODUCTION

According to a study [1] 77% of all electronic failures of cars are caused by software. Detecting these bugs early prevents reputation loss ("car parks, driver walks"), reduces cost and improves time-to-market.

For complex systems like powertrain control, it is of particular importance to understand and analyze the behavior in all possible scenarios thus increasing the software quality. Cycle accurate virtual prototypes of the main components are a perfect environment for stimulation of any use case and for a detailed analysis of the resulting system behavior.

When changing to the 90nm process the costs for the chip architecture development started to exceed the costs for layout for the first time. A virtual prototype allows to do architecture exploration during the concept phase to efficiently select the most suitable IP blocks and reliably dimension system resources based on quantitative measurements.

Software development based on virtual prototypes enables an early development start, improved visibility of internal resources and ease of debugging as well as the ability of generating arbitrary system stimulation. The latter allows to analyze corner cases and to increase the software test coverage.

There are a number of known modeling approaches mostly differing in modeling abstraction levels, the language as well as the tooling. Common to all approaches is the requirement for models that are supposed to be fast, accurate and require only little effort to be developed, verified and maintained. Obviously these are contradicting requirements, however methodology and tools may help to optimize all three requirements at same time. The actual accuracy requirement heavily depends on the use case of the virtual prototype and therefore needs to be carefully selected for the sake of the other requirements.

C/C++- BASED VIRTUAL PROTOTYPES (VP)

C/C++ models of hardware IP have been mostly used for functional and cycle based simulation in the past, mostly known as Instruction Set Simulator (ISS). The introduction of SystemC being a C++ class library has extended the scope to include fully time accurate models offering an event driven simulation paradigm as well as the concept of constructing a system model from component models.

At the functional end the use of Matlab/Simulink and UML is continuing into even more abstract models which rather model applications and algorithms. At the end of full timing accuracy the hardware description languages like VHDL and Verilog are used. Whereas hardware description languages as well as SystemC have been specifically designed to describe and model hardware, C/C++ is a general purpose language offering no specific support for modeling.

The main disadvantage of HDL is the limited simulation performance. This is due to the fundamental computation model that is based on events and processes which enforces the usage of a complex simulation engine. This also applies to SystemC in case the event driven and the process language constructs are being used. C/C++ is a procedural language that enforces the serialization of parallel processes taking place in the hardware when described in the model thus almost eliminating the scheduling overhead at runtime. The cycle based simulation approach offers the most efficient way to synchronize multiple components. Furthermore C/C++ is well known and accepted by the hardware and software designer providing a common language.

VIRTUAL PROTOTYPE APPLICATIONS

Virtual Prototypes are used for:

- Architecture exploration
- Executable specification
- Software development and test
- System analysis and test

ARCHITECTURE EXPLORATION

At the beginning of the design of a complex system the partitioning between software and hardware is not yet determined. Also the specific selection of IP blocks and the system interaction as well as their system performance are mostly unknown and are usually based upon some very high level estimates. In order to explore the system design space based on some quantitative and reliable measurements, a virtual prototype of the IP blocks need be developed. The simulation of the most critical software kernel routines then allows generating meaningful system profiles. Based on these profiles the system can be optimized to reach compliance with the specification without wasting of hardware resources, area as well as power.

In order to be able to design a significant amount of architectural alternatives it is important to start from rather abstract models with only little granularity. The IP block's behavior is modeled only in a purely functional manner where latencies are only approximated by a fixed average value. Instead of modeling bus protocols to the last cycle, the interaction of IP blocks is simplified to read and write accesses accounting fixed latencies only.

The effort of creating such abstract models has to be extremely low, using a standard programming language, strong modeling and interface guidance and automated component framework generation avoiding repeating modeling tasks.

Only after a certain number of design choices have been made, the accuracy of the prototypes needs to be refined in order to provide more accurate system performance measures and providing a higher degree of corner case coverage and overall completeness of the modeled IP's functionality.

Once the refinement of the components of the virtual prototype has been completed and the architecture has been finalized the virtual prototype provides the means of functional correctness as well as of system performance.

EXECUTABLE SPECIFICATION

In the development process of complex systems (e.g. chip design) there is usually a customer and application oriented concept engineering group which specifies all parts of the system. These specifications are implemented by the design team, which creates the microarchitecture and focuses on cost and performance optimizations. The problem is, that native language, used for these specifications, always is ambiguous, leaving room for interpretation. The misinterpretation is sometimes only detected, once the new chip is plugged into the board one or two years later.

In case the concept engineering team provides models as executable specifications of components or even for whole systems to the customer as well as the design team, a significant improvement of the link between requirement and implementation is reached:

1. The customer can check, whether the requirements have been fully understood and considered by the concept engineering. An additional motivation is the fact that the development of (low level) software can be started immediately.
2. The process of developing an executable specification by definition includes a check for consistency and completeness whereas the use of a native language requires additional measures. Even if the model is solely used within the concept team, it will improve the native language specification and avoid iterations with the design team.
3. From the design team's point of view, the model is an unambiguous starting point for the IP implementation. In some cases it can be even used as a golden reference for the verification. An alternative approach is to rerun the functional test cases of the model for the implementation.

SOFTWARE DEVELOPMENT AND TEST

Time to Market is very much impacted by the speed in which software is developed and integrated into the hardware. Without the use of virtual prototypes, the software development and test depends on the availability of evaluation boards or other kinds of physical prototypes which usually become available late in the design cycle. A virtual prototype that has been used already as an executable specification is available much earlier. Even if certain modifications become necessary during hardware development, once the design is frozen, a stable and reliable prototype is available. Due to the fact that software models are able to expose internal resources without any limitations the visibility into the modules can be superior in comparison to hardware. The flexibility of software models also allows adding further debug capabilities, provided that the tool environment offers respective support.

Enabling software development on the basis of a virtual prototype in parallel to hardware design can help in achieving higher quality software in the same time frame.

Virtual prototypes furthermore add the ability to feed specific system level test patterns to a subsystem thus allowing system stimulation and creation of arbitrary scenarios easily.

Due to the compromise between model performance, completeness (also of the environment model) and accuracy, software testing on a model will always be in addition to conventional ways of software engineering based on hardware prototypes.

SYSTEM ANALYSIS AND TEST

The analysis of system level interdependencies is the key for the ability to break up a complex system into smaller, less complex pieces (divide and conquer). Such smaller pieces of the system can then be tested individually using a realistic stimuli and response behavior of the missing part of the system. Ideally complex systems are designed in a hierarchical manner using well defined interfaces between subsystems.

The advantages of dealing with subsystems are:

- Focus on less components and hardware effects
- Higher performance allows to use cycle accurate models
- Stimuli and system response can be used to create arbitrary stimulation and response

PCP/GPTA SUBSYSTEM AS VP EXAMPLE

In order to demonstrate the development and testing of ECU software the PCP subsystem of Infineon's AUDO-NG family is used as an example. The PCP subsystem consists of multiple closely interacting components. The PCP2 is a programmable, interrupt driven processor that handles service requests from all system peripherals independently of the main Infineon TriCore CPU. The General Purpose Timer (GPTA) is a powerful and complex timer peripheral executing the most time-critical tasks. Its sophisticated network of timer cells is programmed via memory mapped registers which are accessed over the System Peripheral Bus (SPB). The SPB is a multi master bus using priority based arbitration producing variable access latencies. Whenever the GPTA has completed one of its parallel running tasks requiring reprogramming, it initiates the execution of service routines by the PCP2. For this purpose it sends service requests to the PCP's arbitrating interrupt system. The interrupt latencies heavily depend on the priorities and duration of higher priority pending services.

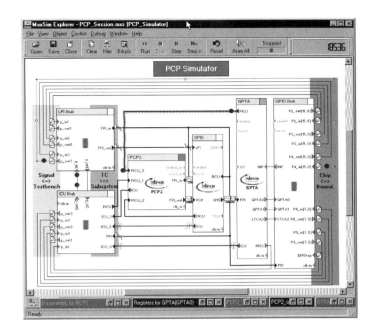

Figure 1 PCP/GPTA Subsystem

The external input and output events observed or generated by the GPTA are multiplexed by the General Purpose Input Output (GPIO) component which connects on-chip and off-chip signals. The interaction between the PCP and TriCore subsystems takes place via the interfaces of the TriCore Interrupt Control Unit (ICU) and the Local Memory Bus to Flexible Peripheral Interconnect Bus Bridge (LFI).

From the software point of view the programming of the PCP2 represents a challenge because of the functional complexity and numerous firm real-time conditions. Due to the multifaceted interdependencies of all involved modules (on-chip and off-chip) testing and debugging on the real hardware can be time intensive, inefficient and error prone, especially when it comes to timing issues. In the sequel details of the software development methodology based on the cycle-accurate virtual prototype of the PCP ECU subsystems are given.

In order to manage the functional complexity it is desirable to first reduce the interdependencies to a minimum, e.g. by simulating the GPTA on its own running as a single task using scripts for the register programming, stimuli generation and checking against expected results. This assumes the availability of a scripting language allowing architectural resource access, stimuli generation as well as testing capabilities. The virtual prototyping tools should also support comfortable monitoring and debugging of the script (Figure 2).

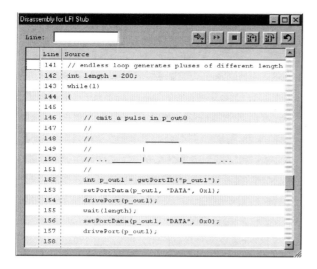

Figure 2 Script Debug View

In order to understand and analyze the behavior of a highly complex component like the GPTA, full visibility of architectural details is the key. The programmer needs to inspect component register subfields, symbolic values and visualize the internal state like clock busses, event output state of timer cells and other hidden states that further complement the understanding of the internal functionality. In the case of large and complex register structures, like in the GPTA, it is helpful to have access to online documentation with hints about the functionality of the bits and the address mapping *(Figure 4)*.

Figure 3 Register View with Online Documentation

In addition, the logic analyzer tool helps to analyze signals and values over time. Advanced execution control using conditional breakpoints on register, fffmemory or signals enhances the ability to efficiently track and locate specific events. The above functionality becomes even more important when dealing with functional interdependencies between parallel software tasks. The programmer needs to deal with the proper use of shared resources concurrently accessed by the tasks. E.g. it must be avoided that shared resources like the clock divider of the clock bus gets accidentally reconfigured or that different tasks route their output to the same pin. Single cycle execution is required to

enable detailed inspection of the system resources on any clock edge.

As the next step, we propose to take the relevant component inter-dependencies into account and therefore gradually increase the scope of the system. Timing related interdependencies mostly deal with the problem that shared resources like the SPB as well as the PCP2 introduce variable latencies depending on the overall system conditions. Furthermore, it has to be ensured that various scenarios of off-chip closed loops are properly handled by the software. The analysis of all such interdependencies requires system level visibility and controllability as well as flexibility to efficiently develop meaningful scenarios. It is of significant benefit to the programmer if the tool offers protocol aware visualization of the component communication and provides a history of transactions using bus monitors *(Figure 4)*.

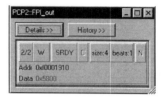

Figure 4 Bus Monitor

Enhanced execution control of the virtual prototype allows the programmer to set breakpoints on communication events between components, thus being able to track cycle by cycle the GPTA interrupt request resulting in a PCP channel program response reconfiguring the GPTA.

If suspicious behavior is detected, it is helpful to analyze the trace information of the current and past input and output signals e.g. produced by the GPTA or GPIO, side-by-side with traces of registers, bus accesses and interrupt requests *(Figure 5)*. In this way the interdependencies can be easily tracked down within a single view window allowing precise measurement of the latencies between events.

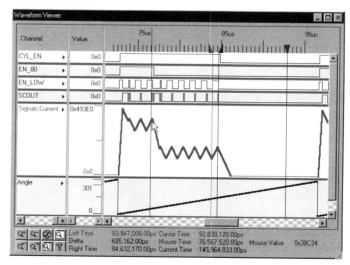

Figure 5 Trace Analyzer

Once a satisfactory basic functionality of the system is reached, the next step is to evaluate the performance of the system as a whole, so further dedicated optimizations can be conducted. In the case of the PCP, the latency information of the SPB bus is an indicator of bus performance and potential bottlenecks which could seriously limit the performance of the system at critical times. In order to generate quantitative bus traffic data and identify worst case conditions, the information about bus transactions needs to be collected and visualized *(Figure 6)*.

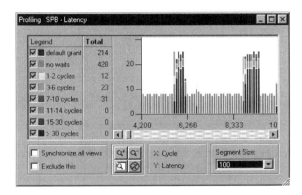

Figure 6 Bus Latency Profiling

The solutions for bus load problems can range from simple changes in bus priority schedule up to the need to restructure the system or make use of a more powerful bus system.

CREATING A NEW MODEL

Before launching any modeling activities the most basic decision has to be made on the simulation environment. This initial decision will determine the long term success of the modeling strategy. Setting up any detail of modeling methodology as well as the simulator environment from scratch requires the highest possible efforts. Mostly this high effort can be brought down to lack in maintenance, documentation and support for semi-professional in-house solutions. If any recommendation towards simulators can be given, then it is to rely on buy-in of methodology (including simulator support) and IP blocks to the most possible extend. Public, license free solutions may seem promising considering the aspect of initial investment, however may suffer from not offering a wide spectrum of features like GUI, visualization widgets and debug capabilities.

The higher the complexity of the component to be modelled, the more important it becomes to interpret the specification for the hardware "first time right". Therefore an ideal team setup has to include software as well as hardware design experts.

If the accuracy of models regarding the aspects functionality and/or timing is of importance no way leads

around verification of models. And as in hardware verification, it is the chosen methodology that will determine the quality of verification results. A verification strategy that runs completely with disregard to the related hardware module cannot deliver sufficient quality. The past has proven, that only equivalence checks at model boundary against the reference hardware module is not enough. In an ideal world, model and hardware verification should share a single verification environment to allow any judgment of accuracy.

GENERAL MODELING CONSIDERATIONS

Modeling methodologies and languages (e.g. C++) impose nearly no limit on what to describe and how to describe it. This section contains some basic considerations how to find the sweet spot between accuracy, completeness, speed and effort.

SPEED VERSUS ACCURACY

For certain types of models, there is a clear trade-off between accuracy and simulation speed. For instance a very straightforward CPU model can be based on a big case statement in ANSI-C, which computes the operands based on the instruction opcode. This model will be very fast and execute the behavior of the CPU functionally correct, however it will not reflect any timing effects due to pipelining, instruction interdependencies, memory latencies, etc. Adding such details in a fully accurate fashion may require to model combinatorial logic, representing every gate by an if-statement, which needs to be reevaluated in each simulation cycle. Obviously such a model will be much slower.

For some CPU cores there is the possibility to have a very fast and efficient model by just "mapping" the instructions of the simulated CPU to appropriate instructions of the simulation host computer. For other types of hardware (e.g. random logic) this type of abstraction is impossible.

The overall simulation speed (frequency f_s) of a system can be approximately calculated from the simulation frequency f_i of the components:

$$f_s = \frac{1}{\sum \frac{1}{f_i}}$$

Obviously a system can never be faster than the slowest component. Therefore only the performance optimization of the slowest components results in significant overall performance improvements.

SYSTEM COMPLEXITY

According to Moore's law the size (number of transistors) of SoCs grows by a factor of 2 every 18 months. An optimistic estimation is, that a system with twice the size needs also twice as much functional evaluation and twice as many tests in terms of clock cycles (doubled system simulation depth in cycles). However a system with twice the size has only half the simulation speed on the same computer. As a conclusion, the system simulation performance requirements grow at least by a factor of 4 every 18 months or 160% per year. The problem is, that the speed of workstations or PCs only increase by roughly 50% every year.

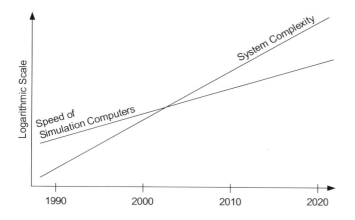

Figure 7 : Simulation Speed Requirements grow faster than the Performance of Simulation Computers

This fundamental trend needs to be taken into account especially for complex systems, when planning a long term simulation strategy. The only way out of this dilemma is to either raise the abstraction level or restrict the scope of the simulation to reasonably sized subsystems, or to do both at once.

CREATION AND VERIFICATION EFFORT

The obvious approach to reduce the modeling and verification effort is to use components from a modeling library. If this is not possible, it needs to be decided for which abstraction and accuracy level the model is to be created. If the IP exists for example as RTL design an RTL to C converter tool could be used, ideally being correct and verified by construction. However such a generated model tends to be nearly as slow as running the RTL description on a HDL-simulator. Another option is the development of a model in a more abstract, less structural and more efficient way and reuse the functional test bench of the RTL design for verification. This greatly reduces the initial verification and thus overall effort, requires however usually a very accurate model.

PUSHING THE LIMITS

Since a model can be considered just as a piece of software, all software performance optimization methods and tricks can be applied. For instance the architecture and the model of computation can be completely different (e.g. functional event based) from the RTL design. Going further the simulator scheduling, structural architecture and functional behavior can be mingled together in the model code or parts can be even written in assembly language. The drawback of these approaches is, that they make the model much harder to create, verify and to maintain and the risk of functional bugs is growing strongly.

On the other hand a model is the more valuable the sooner it becomes available. If the RTL design or even silicon already exists, the benefit of a model is limited. Since the overall speed of a simulation is determined by the slowest component(s), the optimization of one part has only a limited effect. In essence it has to be carefully decided, where to spend effort on speed optimizations.

SUMMARY OF THE MODELING CONSIDERATIONS

The art of a modeling project is to identify in the beginning the sweet spot between accuracy, completeness and speed and to reach this point with the available resources on time.

CONCLUSION

The use of fast and accurate models for architecture exploration, executable specification, software development and system analysis strongly improve the system design cycle in terms of quality, risk reduction and time to market. For maximum benefit the models of the components and the system have to be available as part of the specification phase. This can only be achieved by using a professional, efficient and flexible simulation tool including a broad model library as well as through reuse of existing component models to the greatest extend. The reward is the fact, that the software will be already available, once the first silicon comes back from the fab and having it up and running in no time.

REFERENCES

1. G. Jacobi, "Software ist im Auto ein Knackpunkt", VDI Nachrichten, 28th February 2003.
2. Snapshots: RealView ESL Tools, MaxSim Explorer by ARM

CONTACT

Joachim Krech
Development Systems (ESL)
Tel: +49 (0) 2407 908620
Email: Joachim.Krech@arm.com

Albrecht Mayer
Principal Advanced Emulation Concepts
Tel: +49 (0) 89 234 83267
Email: albrecht.mayer@infineon.com

Gerlinde Raab
Staff Engineer
Tel: +49 (0) 89 234 87166
Email: gerlinde.raab@infineon.com

DEFINITIONS, ACRONYMS, ABBREVIATIONS

HDL: Hardware description language

ISS: Instruction set simulator

RTL: Register Transfer Language

VP: Virtual Prototype

ANSI: American National Standards Institute

To Test the Need and the Need to Test- Testing the Smart Controller Network for the Chassis of Tomorrow -

Harald Deiss, Horst Krimmel and Oliver Maschmann
ZF Friedrichshafen AG

ABSTRACT

Hardware-in-the-loop (HIL) simulation has become a key technique for the validation of today's automotive electronics. OEMs and suppliers are investing heavily in hardware-in-the-loop equipment and tests. Typically, suppliers test the electronic control unit (ECU) as a component. The OEM on the other hand tests the ECU more from a network point of view. This paper describes the main differences between component and network HIL tests.

ZF Friedrichshafen AG has been using HIL test benches since 1985. In order to ensure high quality, especially with respect to network aspects, we not only test the ECUs as components but as part of the network. For that purpose, and to stay on the leading edge of HIL technology, ZF has set up a new test bench for networked HIL testing. The control network contains the driveline and chassis domain. The devices tested are, e.g., automatic transmission, torque on demand transfer case, torque vectoring axle drive, electric power steering, active steering system, active stabilizers, variable dampers, levelling control as well as a brake control system (ESP). This article presents the key innovations of this network test bench.

As tests are typically dispersed along the multistage supply chain between the vehicle manufacturer and several suppliers, cross-organizational test management is needed. This is the only way to ensure that all necessary tests are carried out, no tests are omitted and no redundant tests are unnecessarily performed.

Reuse is the key to efficiency. ZF Friedrichshafen employs a two-step reuse strategy which distinguishes between reuse of complete test cases and reuse of elementary test modules. This paper describes our reuse concept for test cases and elementary test modules on the basis of test libraries.

INTRODUCTION

For many years, hardware-in-the-loop testing has been a major instrument for testing automotive electronics. Currently, OEMs and suppliers are investing heavily in hardware-in-the-loop equipment and tests. Is there enough benefit to justify this expenditure? In order to determine the need for HIL tests, we test their efficiency by means of examples. Today, at least three trends are increasing the need for tests and requiring optimised hardware-in-the-loop technologies and processes:

- Firstly, the constantly increasing complexity of networked functionality (e.g. recent hybrid driveline and energy management) is calling for optimized testing methods. This is true for single electronic control units (ECU) as well as a network of many ECUs. See Fig. 1 and Ref. [1]
- Secondly, diversification in the vehicle market has resulted in more and more vehicle classes, including many cross-over vehicles. See Fig. 2. At the same time, there is a trend towards fewer vehicle platforms. Therefore, even if the ECU hardware is an off-the-shelf component, ECU software comes in many variants to permit sufficient differentiation.
- Thirdly, development and testing are increasingly distributed among the vehicle manufacturer and several suppliers in a multistage supply chain. Cross-organizational test management is therefore required.

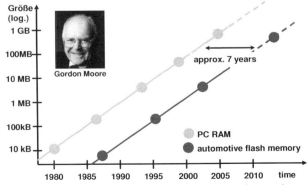

Fig. 1: Increase in complexity of automotive electronics

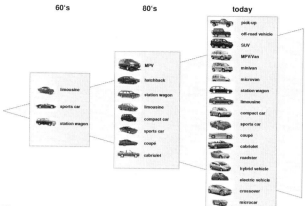

Fig. 2: Emergence of many variants

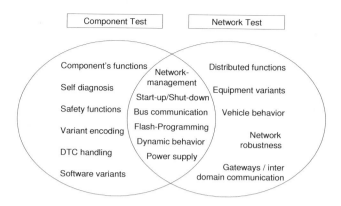

Fig. 3: Main testing fields of component and network test.

The next chapter shows exemplarily the essential need for HIL technology for component as well as network testing. It also provides a description of the main differences between component and network test benches.

The third chapter presents ZF's new HIL test bench for networked driveline, steering, and chassis systems. New technologies are illustrated by means of exemplary solutions proven in driveline, steering and chassis development at ZF Friedrichshafen. This domain is especially suited because of its high safety and reliability requirements (Fig. 6).

The fourth and the fifth chapter deal with the cross-organizational test management requirement and the efficient reuse of HIL tests. And we discuss the recent progress in hardware-in-the-loop technology allowing management of the increasing complexity as well as the rising number of software variants.

COMPONENT TEST VERSUS NETWORK TEST

In the past, the vehicle was unfortunately often used for the first integration test. As a consequence, additional development loops arose and quality became worse. Now, both OEMs and suppliers perform HIL tests, but from quite different points of view.

- Suppliers test the ECU as a component. These tests concern the component's self-diagnosis and functionality, etc. See Fig. 3 for details. Ref. [2, 3].
- The OEM tests the ECU from a network point of view in order to prove correct ECU communication, network management, and flash behaviour in the network, etc. See Fig. 3 for details. Ref. [4].

EXAMPLE FOR COMPONENT TEST

A supply voltage test is a typical example for a test at the component level. During the test run, different battery voltage curves are fed through by the system. The relevant voltage thresholds and robustness requirements are laid down in the control unit specifications. From this, the expected control unit behaviour can be derived and inspected.

- The system may only shut down below a specified operating voltage.
- The system must recover automatically within a specified time frame.
- There must be exactly one (e.g. undervoltage) error entry in the error memory, and no subsequent errors.
- Unwanted system response (e.g. undesired actuation of the system) may not occur at any time.

EXAMPLE FOR NETWORKED TEST

Fig. 3 shows a big overlap of component and network test. This means many features are tested both at the component and the network test bench. One reason is that the network test bench is more realistic than a residual bus simulation used at the component test bench.

As an example at network test level we discuss the test of distributed vehicle dynamics control. The stability control is using brake interventions as well as automatic steering support (active steering). During an oversteer driving maneuver the stabilizing begins with automatic steering support. Only if the vehicle is not stabilized in this way the brake intervention is done in the second phase. This test contains several steps.

- At first, we connect all ECUs with relays to the supply voltage (e.g. KL30) and the bus. The ECUs are still "down" at this moment. With ignition signal (e.g. KL15) the ECUs start-up. Then the engine starts. The vehicle has zero speed.
- A double lane change driving manoeuvre is started. The speed is chosen high enough to achieve both an automatic steering support and a brake intervention during the driving manoeuvre.
- We test if the steering support is coming first, with the correct direction and size.
- If the difference of desired and actual yaw rate is still increasing, we check if brake intervention is coming in the next phase and if the intervention is correct.
- At the end of the test, we have ignition off, the ECUs are shut down via network management. Finally the ECUs are disconnected from power supply. The test bench is in its initial state again.

TO TEST THE NEED

In view of the magnitude of the effort, the need for HIL testing is often questioned. We analysed four projects with respect to the efficiency of HIL testing. We related the number of identified faults (as a measure of success) to the number of test cases (as a measure of effort). We found ratios in the range of 3 to 30 percent (Fig. 4).

As to the projects showing a high ratio, many test cases were available in early project phases already. Many errors were found in the lab and not in the car. In practice, the test specification and, as a consequence, the HIL tests are often available in late development phases only. This is because in early development phases most developers are involved in implementing the system's functionality. As soon as the functionality is implemented, the developers have enough time to write the test requirements. The projects with a low ratio belong to this scenario. The benefits that remain in that case consist in the documented validation of ECU and network and the possibility to make regression tests for new releases.

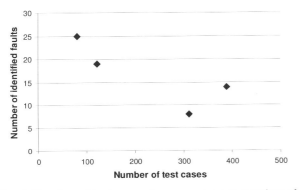

Fig. 4: Number of test cases in comparison to number of identified faults (four typical projects).

ZF'S NETWORKED HIL TEST BENCH

ZF Friedrichshafen AG is a leading supplier of driveline, steering, and chassis systems. To ensure high quality, especially with respect to network aspects of the electronic control units, we not only test the components but the entire network. For that purpose, a new test bench for networked HIL tests has been set up (Fig. 5). Most of the driveline, steering and chassis systems are part of this test bench. Fig. 6 shows some of the tested systems.

- Automatic transmission (AT)
- Torque-on-demand all-wheel drive (VTG)
- Torque-vectoring axle drive (TV-HAG)
- Hybrid drive
- Electric power steering (EPS)
- Active steering (AS)
- Variable dampers (CDC)
- Active roll stabilisation (ARS)
- Air suspension levelling control (1ALF)
- Brake, Electronic Stability Program (ESP)

Fig. 5: ZF's HIL test bench for ECU network testing

Today several tool suppliers provide all-in-one solutions for HIL tests. ZF Friedrichshafen AG nevertheless decided to realise a completely modular solution (Fig. 7) in order to be able to

- combine best-in-class components from different suppliers, e.g. virtual driver, vehicle model, FlexRay equipment, etc.;
- use specialized in-house solutions if necessary from a know-how point of view, if cheaper, or reasonable in other respects;
- benefit from maximum flexibility with respect to future innovations in the field. For instance test automation or test management tools.

HARDWARE AND SOFTWARE BACKBONE

The hardware backbone of the test bench is a dSPACE real- time environment. Every three systems share one processor main board. The vehicle model is run on a separate main processor board for computing capacity reasons. All together, we have five processor main boards which are all connected via high-speed optical GigaLink bus.

Matlab/Simulink is the integration platform for hardware control (dSPACE real-time interface RTI), vehicle model, virtual driver, road, etc. We use dSPACE tool chain to control, automate and visualize the test bench.

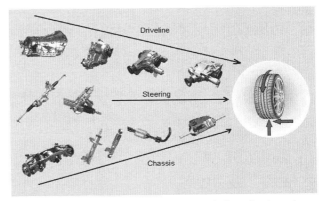

Fig. 6: Systems tested in driveline and chassis domain

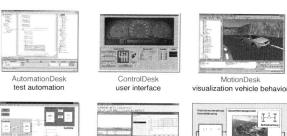

| AutomationDesk test automation | ControlDesk user interface | MotionDesk visualization vehicle behavior |
| Matlab / Simulink integration platform | Canalyzer bus monitor | Simpack / Dymola / C-Code vehicle model |

Fig. 7: Test environment

VEHICLE MODEL

A real-time vehicle model is necessary for the calculation of realistic stimuli for the ECUs. As we need highly accurate driveline and chassis parameters, we have specialized models.

Chassis: Elastokinematic models based on characteristic curves are usually sufficient for HIL applications. In order to have a seamless tool chain from offline to online simulation, we use a multibody system tool (Simpack) permitting C-Code export. This C-Code model is then integrated as Simulink S-function. Code export from multi-body system tools and use of the C-Code on real-time computers constitute a real challenge. We spent a lot of time on that, especially on optimising the computation effort of the model. Now we have a processor load of approximately 80 per cent for the chassis model (ds1005 main processor board). We achieve good numerical stability of the complete vehicle by integrating chassis, tyre, and driveline model by means of one separate integrator. In parallel, Simulink works with Euler forward algorithm with 1 ms fixed step size.

Driveline: We use a highly run-time-optimised in-house driveline model that takes 5 per cent processor load only. The model is optimized regarding stick / slip transitions. Each shaft of the transmission and the driveline is represented by its speed and inertia. Transmission control and brake system hydraulics are also included.

virtual driver, virtual road vehicle model (driveline, chassis, hydraulics)						
GigaLink optical high speed bus connecting the 4 racks						
real time processor board (ds1005)						
real time I/O board (ds2211)						
actuators sensors	actuators sensors	actuators sensors				
signal switchbox	signal switchbox	signal switchbox	1st rack	2nd rack	3rd rack	4th rack
AT-ECU	VTG-ECU	Hybrid-ECU		TV-HAQ-ECU HAQ-ECU	ESP-ECU EPS-ECU AS-ECU	CDC-ECU ARS-ECU 1ALF-ECU
power switchbox	power switchbox	power switchbox				
power supply	power supply	power supply				
ECU's residual bus simulation (CAN; FlexRay, ...)						

Fig. 8: Block diagram of the test bench arrangement

MODEL VALIDATION

Realistic simulation requires an accurate vehicle model. Approximately one man-year had to be invested for model validation. This includes collecting model parameters, measuring unknown vehicle parameters, and comparison of simulation results with vehicle measurements.

We optimised the vehicle model concerning longitudinal and lateral dynamics. Values concerning driving dynamics are reproduced exactly in the linear range. In the limit range of driving dynamics, values like transversal acceleration are calculated with an exactness of approx. 1m/s². The braking distance can be simulated with an accuracy of approx. 1 m for full brake application from 100 km/h with engagement of slip control. This accuracy is sufficient for most test objectives. See Fig. 9 for an example of validation precision.

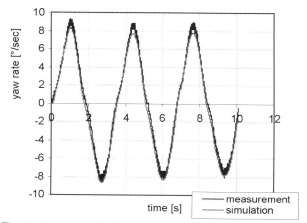

Fig. 9: Model validation, comparison of yaw rate during sine steering with 0.6 Hz

FAULT SIMULATION

The test bench offers all state-of-the-art options for fault simulation:

- Each ECU has its own independent voltage supply. This permits simulation of disturbances in the main power supply or ground offset effects.
- There is a relay box to produce defective lines or contact problems in the electrical connections of the ECU (e.g. short circuit, etc.).
- Last but not least, the CAN communication of any ECU can be manipulated. CAN messages can be added or removed. Or the data content of any CAN message can be manipulated. For this purpose, each ECU can automatically be separated from the communication network by a gateway. Furthermore, stress testing is possible by increasing the bus load through "residual bus" simulation.

FLEXRAY

The test bench offers flexray infrastructure for those systems (like variable dampers) that already use flexray communication.

LINKING THE REQUIREMENTS PROCESS TO THE TEST PROCESS

To make sure that each software requirement is thoroughly tested, we link the requirements process to the test process. The system, software and test requirements are therefore all linked up in the same requirements management tool (DOORS).

The complexity resulting from such linking should not be underestimated. For a typical automotive project has several hundreds of requirements at each level: system, software, and testing (Fig. 10). Despite the help of the tool chain, it is not easy to keep all three levels (system, software, testing) consistent and up-to-date.

- Since the requirements process is distributed among several companies, cross-organisational management involving the OEM, the 1st tier supplier and several 2nd tier suppliers is complex. To solve this task, we use the "two-module strategy" of DOORS, which allows asynchronous many-sided modification of the data base and subsequent resynchronisation of all parts. See Fig. 11 for more details.
- Automated tests have so far not been included in the requirements management tool at ZF Friedrichshafen. The tool chain between test requirements and tests is consequently broken. Each test requirement in DOORS contains a reference number to the test in order to ensure traceability. ZF is working on closing the tool chain to achieve seamless referencing from system requirement to each test case.

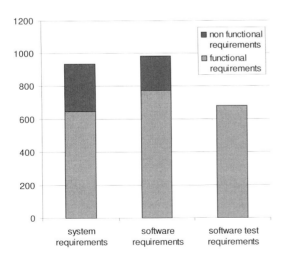

Fig. 10: Example: Number of requirements at system, software and software test level (typical project).

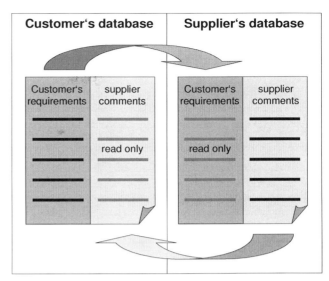

Fig. 11: Tool supported requirement management process. Exchange customer's requirements and supplier's comments.

REUSE OF TESTS

Reuse is the key to efficiency. If complete test cases can be reused in a new project, this is especially efficient. Unfortunately, reuse of complete test cases works in the majority of cases only within the same product line (supplier's view) and / or vehicle platform (OEM's view). ZF therefore uses a two-step reuse strategy and distinguishes between reuse of complete test cases and reuse of elementary test modules.

REUSE OF COMPLETE TEST CASES

We analysed the reusability of complete test cases from a supplier's point of view by means of an example, Ref. [5]. The example refers to the HIL test of an active steering system which provides a variable steering ratio and automatic steering support for vehicle dynamics stabilisation (Fig. 6). The analysis contains all HIL tests conducted by the supplier. The tests performed by the OEM are not considered. As a constraint, we assumed that reuse happens in the same testing (tool) environment. Analysing the reusability of the HIL test (Fig. 12) we found:

1. 43 % of the tests are product-specific and - with few modifications - reusable for vehicles of the same or a different OEM. Example: High-level system functionality, such as driver assistance features of a steering system.
2. 29 % of the tests are product-specific and reusable for vehicles of the same OEM. Example: Supply voltage tests according to the OEM's corporate standard.
3. 8 % of the tests are system-specific and not reusable at all.

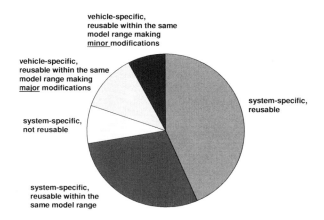

Fig. 12: Example: Reusability of test cases. (Estimated, typical chassis project)

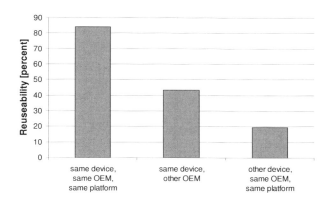

Fig. 13: Example: Reusability of test cases (estimated, typical chassis project).

4. 8 % or 12 % of the tests are vehicle-specific and reusable within the same vehicle platform, with minor or major changes being required. Examples of this set are tests concerning communication, diagnosis services, flash download, etc., which are typically the same within a vehicle platform.

Analysing the same data the other way around we can draw the following conclusions (Fig. 13): If tests for a specific system are transferred

*   to another vehicle in the same platform, then the level of reuse (with minor and major changes) is about 90 %, since the unit-related as well as the vehicle-related requirements are largely the same (1. + 2. + 4. item above);
*   to a vehicle of a different OEM, then the level of reuse drops to about 40 per cent (1.);
*   to a different unit in a vehicle of the same OEM, then the level of reuse is also about 20 per cent (4.);
*   to a different device in a vehicle of a different range or platform, then level of reuse depends heavily on the project parameters and is mostly far below 30 per cent.

Taking into consideration that a project mostly involves several hundred test cases, and that on average a tester automates one test case per day (as a rule-of-thumb), then even a level of reuse of 30 per cent will permit significant savings in terms of funds and capacity.

REUSE OF ELEMENTARY TEST MODULES

The reuse of elementary test modules is much easier than the reuse of complete test cases. We define an elementary test module as being

*   a small fraction of a complete test case;
*   generic; and
*   usable in many test cases.

Hence, these elementary test modules are characterized by their reusability. Examples for such elementary test modules are blocks for adjusting stimuli and checking test parameters, as well as start-up of the vehicle, running through different terminal states, right up to complete driving manoeuvres.

TEST LIBRARY

ZF Friedrichshafen is organising the reuse of tests by means of a test library. Analogous to the discussion in the previous chapters, we distinguish between an "elementary test module library" and a "test case/scenario library".

*   Elementary Test Module Library: The module library implemented by ZF includes approximately 400 elementary test modules. These test modules are arranged in 10 groups (Fig. 14). Using these modules makes it possible to create new test scenarios via "drag and drop", quickly and simply.
*   Test Scenario Library: This library comprises a collection of test specifications and the corresponding test scenarios, which are used in different projects. It is therefore not necessary, for example, to reprogram the diagnostic tests for each project. Tests can be used for several different control units. The test scenario library also includes system functions the driver can experience, e.g. ESP intervention to stabilize the vehicle during braking.

CONDITIONS FOR EFFICIENT REUSE

Analysing many projects regarding the reuse of HIL tests at ZF Friedrichshafen, we ascertained the following key factors for efficient reuse:

*   A hardware abstraction layer is necessary for easy transfer of a test case from one test bench to another. In the test case you refer variables which are link to the test bench hardware by a mapping file.
*   Separation of test procedure and test data: The separation of test procedure and test data is an essential aspect in reusing and configuring test scenarios. This means that different specifications can be considered through modification of the parameters without the need to modify the test scenario itself.

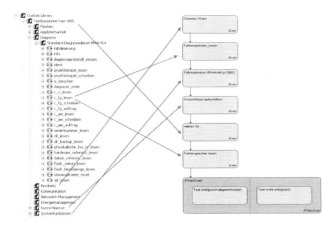

Fig. 14: ZF's test library. Build up new test cases by just "plug and play".

- Handling of versions and variants: It is also essential for efficient handling of test scenarios that the language describing the test allows simple handling of different versions and variants. Easy integration of test scenarios into configuration management, including features such as version comparison, branching, and merging, is mandatory - especially for graphic environments. According to our experience, graphic test automation tools still have to be improved for this goal to be met. In projects with many variants we use python script language for test automation. In this way the additional test steps which belong to a variant can be switched on easily.

CONCLUSION

One indispensable prerequisite to successfully secure the system at the test benches is a requirement process that is implemented right from the start of the project. The system specifications and the test specification must be linked to each other and kept up to date throughout the entire development process. This is a real challenge to both OEMs and suppliers.

Despite the great progress that has been made with HIL tests, many errors occur when integration into the vehicle takes place. This is an incentive to find more errors in the laboratory already. The key to success is to have the HIL tests available in early project phases already. If HIL tests are implemented in late project phases, efficiency will decrease.

Reuse of tests is an important means to reduce development costs and increase quality. Its potential heavily depends on the type of reuse. Generally, its potential is underestimated. To raise its potential, many organizations are reinforcing the teams for hardware-in-the-the-loop tests or, if not yet installed, such teams are newly established.

Significant progress in HIL technology would be made especially by standardizing tool interfaces and thus achieving a platform-spanning exchangeability of test scenarios [6].

REFERENCES

1. Krimmel, H; Deiss, H; Runge, W; Schürr, H.: Electronic Networking of Driveline and Chassis. ATZ 108 (2006), Nr. 5
2. Dornseiff, M.; Stakl, M.; Sieger, M.; Sax, E.: Durchgängige Testmethoden für kompakte Steuersysteme – Optimierung der Prüftiefe durch effiziente Testprozesse, Elektronik im Kraftfahrzeug, Baden-Baden, 2001
3. Bauer, G; Tapia, C; Dornseiff, M.: Software teststrategies with increased efficiency for the development of car automatic transmission functions based on the example of a controlled converter clutch, Elektronik im Kraftfahrzeug, Baden-Baden, 2003
4. Waltermann, H.; Schütte, K.; Diekstall, J.: Hardwarein-the-Loop-Test verteilter Kfz-Systeme. ATZ 106 (2004), Nr. 5
5. Krimmel, H; Maschmann, O; Seidt, S; Vogt, D.: Opportunities and Limits of HIL-Tests, ATZelektronik, 2006, Nr. 4
6. //www.asam.net/

A Systems Engineering Approach to Verification of Distributed Body Control Applications Development	2010-01-2328 Published 10/19/2010

Jinming Yang, Jason Bauman and Al Beydoun
Lear Corporation

ABSTRACT

An effective methodology for design verification and product validation is always a key to high quality products. As many body control applications are currently implemented across multiple ECUs distributed on one or more vehicle networks, verification and validation of vehicle-level user functions will require availability of both the vehicle networks and multiple ECUs involved in the implementation of the user functions. While the ECUs are usually developed by different suppliers and vehicle networks' infrastructure and communication protocols are normally maintained and developed by the OEM, each supplier will be faced with a similar challenge - the ECU being developed cannot be fully verified and tested until all other ECUs and their communication networks are available in the final development stage. In such cases, many design and implementation errors associated with each ECU and their interactive functions cannot be identified prior to vehicle-level integration testing, at which time cost of fixing errors would be high for each supplier involved. The errors that are not discovered during integration testing will consequently affect product quality and timely delivery. Even if all the ECUs are available and work for their "happy paths", it will still be challenging to validate the ECU's capability of handling fault conditions. Therefore, a fault insertion testing strategy is essential to fully meet customer's expectations and robust design.

This paper describes a methodology for developing body control applications based on the concept of executable specification, plant modeling, test case generation using various means, and migration of test cases in the virtual test harness model to ECU-in-the-loop testing environment. Unique aspects of the plant modeling, test case development strategies and their value are discussed in this context. For example, behavior models of other ECUs on the vehicle network, fault conditions, and commands from an external diagnostic device into the plant models are incorporated to enable verification of distributed body control applications. For verification, the use of Stateflow for test case development and test case reuse for both requirements verification and ECU in-the-loop testing are also discussed. The methodology described in this paper has been successfully applied to production projects.

INTRODUCTION

Automotive body-control functions of a modern vehicle are nowadays implemented with multiple ECUs that are logically interconnected through one or more vehicle communication networks [1]. As reducing the number of ECUs utilized in a vehicle is a goal, it is becoming a trend to integrate and implement more user functions into one ECU [2], thanks to powerful multi-core CPUs and large memory space in today's microcontrollers. On the other hand, there also exists a trend where an increasing number of ECUs are employed, due to the need of decomposing vehicle-level user functions into multiple functional blocks which are thus to be implemented and distributed across multiple ECUs. Such a need may be originated from both technical and business perspectives including modular architectural design philosophy, visibility requirement for interactive signals among ECUs, reusability of ECUs for different product lines, protections for OEMs' IPs, vehicle architectural design flexibility, etc. Therefore, the industry sees such two opposite trends in vehicle architectural design, which may be termed "centralized design" and "distributed design".

While the total number of body-control ECUs utilized in a vehicle does not significantly go up or go down due to the confluence of two opposite driving forces behind the two trends, the force driving the number to increase appears to be more dominant. The only trend that is definitive here is that the complexity of interactions among the ECUs on vehicle networks has significantly increased in recent years [3].

Moreover, the distributed design approach requires security checks and handshakes between ECUs which further adds complexity to the vehicle system. Such a trend in vehicle system design presents new technical challenges for the verification and validation processes. As a result, to verify the designs and validate the user functions, engineers will need all the ECUs involved in implementing the user functions to be available. While most of the ECUs would be unavailable during the design and implementation phases because most of them are designed and implemented by different suppliers [4], each ECU supplier would be faced with similar difficulties in identifying issues existent in functional design and interactive behaviors among the ECUs and/or subsystems. Consequently, any potential design and implementation issues that should have been identified in the individual ECU development stage would have to be identified and resolved in the vehicle validation and integration stages [5]. Undoubtedly, the failure of early identification of design and integration issues will not only make the development cost much higher but also inevitably affect product quality and timely delivery.

Another dilemma in this respect lies in design verification and validation of fault handling capability [6]. Due to the distributed functions across the vehicle networks, each ECU will have to be capable of handling faults caused by race conditions, deadlocks, missing messages, etc [7]. Once again, without the availability of its surrounding and workable ECUs, the vehicle-level design and fault handling capability cannot be fully verified in the early development stages. More importantly, even in the final vehicle-level validation and integration stages, certain fault conditions may not be easy and/or possible to be created with the ECUs (that most likely function properly and without faults), therefore validating the fault handling capability still remains a challenge. After all, it is always crucial to fully validate the exception handling capability, which is the key to vehicle reliability and safety.

In this paper, a model-based methodology to address the issues and challenges described above will be presented. The methods include creating test cases in various development stages, using plant models[1] [8] to run model-in-the-loop simulations, inserting faults to plant models, reusing the test cases for hardware-in-the-loop (HIL) testing and validation, and dynamically deploying test cases on the HIL tester. The novelty in this work exits in design verification and product validation for the distributed type of body-control ECUs by leveraging plant models and the fault insertion method. The use of virtual test harness models for test case creation and reuse are also emphasized in this paper. Other unique aspects discussed in this paper include the following:

• Generate test vectors by collecting simulated data from virtual test harness models. Such test vectors are then converted into reusable test cases.

• Add fault controls to plant models in order to create fault conditions to the ECU(s) being developed and tested.

• Reuse the plant models for ECU in-the-loop testing by importing plant models to the HIL tester. The test cases created in the virtual test harness environment can be reused for the HIL testing.

• Resolve test sequence and dependency issues by using Stateflow model to do both test vector generation and results checking.

• Project management aspects such as number of reusable test cases, creating test cases for requirement changes, plant modeling as a new activity in model-based design process, tackling issues for design changes and design variants, and seeking customer support for plant modeling.

In this work, the MathWorks' toolset including Simulink®/ Stateflow® and MATLAB programming language was used. The methodology introduced in this paper has been successfully applied to production projects.

A CASE STUDY OF DISTRIBUTED USER FUNCTIONS

As stated earlier in this context, there exist valid reasons including both technical reasons and business reasons why user functions are implemented across multiple ECUs. Here the ***passive start*** function[2] is taken as an example for the said vehicle user function which has become an important user-convenience feature in recent years. In general, when the driver presses the vehicle start button trying to start the vehicle, the BCM will send a signal to the ECU responsible for the *passive functions* (i.e., passive entry and passive start or PEPS). The PEPS ECU will then trigger another ECU to transmit wireless LF signals to the vicinity where the passive key (i.e., a remote keyless entry device) can be located. After the passive key receives the valid challenges, it sends back RF signals to the wireless receiver ECU. If the passive key is authenticated per its encrypted/decrypted algorithm, the BCM initiates the authentication with the power-train ECU. At this time, the BCM ECU may check additional preconditions required for passive start from other ECUs including transmission ECU, the cluster instrument ECU, or other ECUs possessing relevant information. Therefore, the passive

---

[1]The terms "plant" and "controller" primarily employed in control theory for the closed-loop type of control systems have been adopted in model-based design to refer to surrounding systems that interact with the system under developed in a non-strict sense. Both "plant" and "plant models" are used in the references.

[2]Note that all the vehicle functions described here are general information and knowledge which is available in public domains and special care has been taken to ensure no proprietary information to be disclosed in this context.

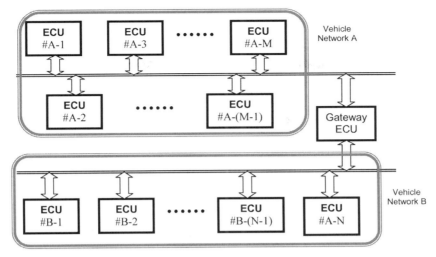

Figure 1. A typical vehicle body-control architecture with multiple ECUs across two vehicle networks with M ECUs on the Network A and N ECUs on the Network B.

start user function may involve the BCM ECU, PEPS ECU, receiver ECU, power-train ECU, transmission ECU, cluster ECU, and more. In a typical vehicle with the passive start function, the feature normally involves sub-functions across an average of 5 ECUs, or up to 40 unit models distributed on multiple ECUs from the modeling perspective. The unit models mentioned here that were created according to the adopted modeling guideline, consist of a set of Simulink blocks or Stateflow models with a defined manageable complexity which may generate 50 to 300 lines of code (excluding lines of comments).

Besides the passive start function, the BCM may also need to include many other functions such as remote start, locking, wipers, lighting, alarming and warning, battery management, etc. Each of these functions may involve other ECUs on the vehicle network(s) and further complicates the system design for the ECU being developed.

The above-described passive start function can be generalized into a generic body-control architectural design as shown in Figure 1. Assume the two vehicle networks are gatewayed by an ECU (named Gateway ECU here) with *M* ECUs on the Network A and *N* ECUs on the Network B. The network A has ECUs labeled ECU #A-1, ECU#A-2, …, ECU #A-M, while the *N* ECUs on the Network B include ECU #B -1, ECU#B-2, …., ECU #B-N. If the ECU labeled "ECU #A-1" is to be developed and the user functions A, B and C are to be validated, there will be multiple ECUs that interact with the one to be developed. In this example, the ECUs needed for a full verification and validation will include ECU #A-2, ECU #A-3, ECU #A-(M-1), Gateway ECU, and ECU #B-(N-1). This is shown in Figure 2.

TEST CASE CREATION IN DIFFERENT DEVELOPMENT STAGES

The method for test case creation is to create test cases in various development stages and then reuse them as much as possible in all validation phases throughout the entire product development cycle. The test cases created in different development stages are intended to test against different aspects of requirements and design. Specifically, the unit-level test cases are intended to verify the inside functional model. Note in this context the term "unit" is used to refer to a scope of a MATLAB model (models) which has a defined manageable complexity and implements a conceptually dividable or a function-wise stand-alone logics and/or algorithms. And a feature model (or a feature) contains a set of the unit models that are integrated to implement a vehicle function. The test cases developed by taking a feature model as a whole are intended to verify whether the unit models can properly work together as required. Using the feature model simulated in-the-loop with *other* feature models (which are usually termed as "plant models"), one can develop test cases meant to identify the interactive issues between/among them. Test cases can also be developed by focusing on a given user function by adding its surrounding functional blocks which will be models either within the same ECU or other ECUs on vehicle networks.

CREATE UNIT/FEATURE -LEVEL TEST CASES DURING MODEL DESIGN

The test cases for unit-level and/or feature-level models are developed using the executable test harness model as described in [9]. The unit-level or feature-level test cases are normally created by manually creating signals in Signal Builder, running simulations and verifying the simulation

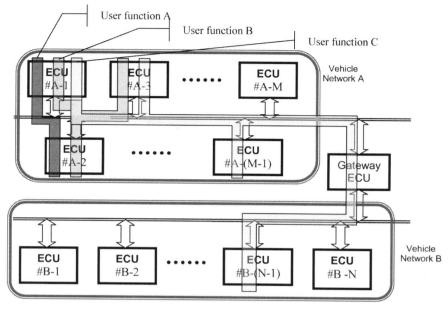

Figure 2. Many user functions are implemented across multiple ECUs distributed on vehicle networks. In this example the User functions A, B and C are all implemented on multiple ECUs.

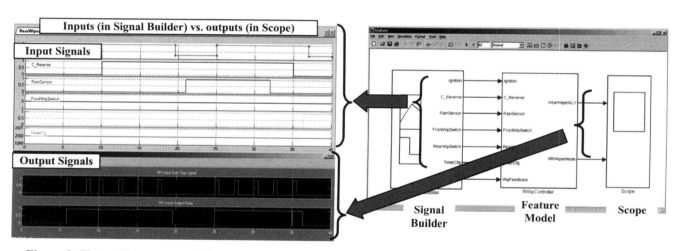

Figure 3. The test harness model is an executable model that can be used to create test cases for design verification. requirement analysis and clarification.

results. The Signal Builder test cases are developed during two activities, (a) when the designers try to verify the design against the requirements, and (b) the phase of requirements confirmation and clarification [10].

The output signals generated from simulation are normally examined and reviewed with development team as well as with the customer. All verified outputs are then stored in the Signal Builder for regression tests and exported in the Excel format for reuse in later validation tests. Note as per the model-based process adopted in this context, the tasks such as storing the simulation results, regression test, and exporting test cases are automated and streamlined by using the

proprietary MATLAB scripts as described in our previous work [9] but it is again illustrated in Figure 3. The so-called "feature model" in the figure can be either a unit-level model or a feature model that encapsulates multiple unit-level models.

CREATE MORE TEST CASES WITH PLANT MODEL "IN-THE-LOOP"

When a plant model is placed "in-the-loop" with the feature model as shown in Figure 4, the model-in-the-loop behaviors can be simulated. In such a closed-loop type of simulation, the outputs generated by the feature model are used to drive

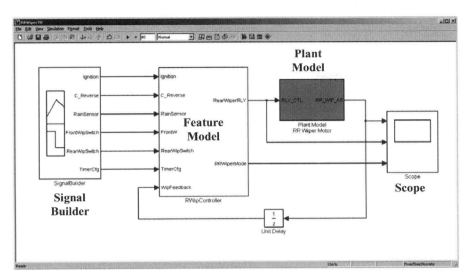

Figure 4. The plant model is placed "in-the-loop" where it is driven by the output of the feature model while its output is fed back as inputs to the feature model.

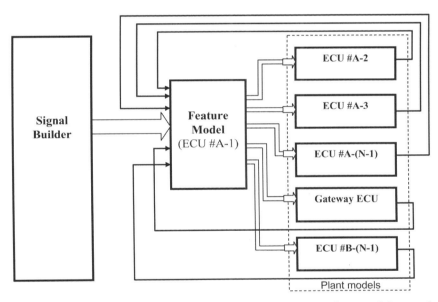

Figure 5. Multiple plant models may be placed "in-the-loop". In the case study, 5 plant models are placed in-the-loop with the feature model.

the plant model, which in turn yields output that is fed back as inputs to the feature model. In Figure 4, one will notice that the number of inputs from the signal builder has been reduced as compared to what is shown in Figure 3 because the input is generated by the plant model. As the simulation runs in-the-loop, more test cases are created and some of test cases created earlier will no longer make sense.

PLACE MULTIPLE PLANT MODELS IN A TEST HARNESS MODEL

Note that multiple plant models may be placed "in-the-loop" in a similar way. If using the five plant models in the example illustrated in Figures 1 and 2, the test harness model will look like what is shown in Figure 5. Note that the topology in the test harness model is different than what is shown in Figure 2 when the ECUs are placed on the vehicle networks, because of the focus on the application layers of software. That is, in the test harness model, all the lower layers of software for the ECUs are omitted for the sake of being able to simulate the

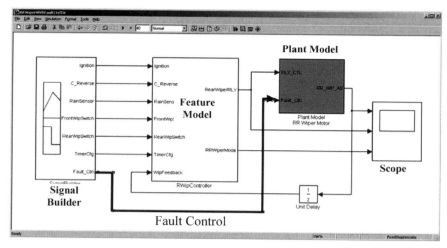

Figure 6. The fault control signal(s) are created in the Signal Builder which directly control the fault conditions for the plant model and indirectly control the inputs to the feature model.

ECUs in the closed loops. Similar to the previous case with only one plant model being in-the-loop, the number of inputs has been further reduced and some of the inputs are replaced by the outputs from the plant models. Therefore, the test cases created in this test harness will be closer to system-level test cases.

Note that the test cases created here may have lower coverage rate for the feature model, however the test cases are more meaningful to reflect the interactive aspects among the ECUs as well as potential issues associated with the service and driver layers when they are ported to hardware in-the-loop testing and validation.

FAULT INSERTION USING PLANT MODELS

As described earlier, it is both necessary and beneficial to create fault conditions using plant models. The fault control to the plant models can be added in order to simulate the fault cases for the feature model. When the plant model is simulated in-the-loop and fault control is added, more test cases can be added. Note there will be more implications to design verification and ECU validation for the following reasons:

• Even if the other ECUs are available in the final validation phase, fault conditions from other ECUs cannot be easily created (as the ECUs most likely operate properly), so the capability of handling fault conditions cannot be validated without simulations in the virtual test harness environment.

• Besides fault conditions, race conditions from other ECUs are not easily created.

• It is often challenging to duplicate issues found in pre-production - manipulating plant models by controlling their fault conditions makes it possible/easier to duplicate specific issues.

ADD FAULT CONTROLS TO A PLANT MODEL

The idea is shown in Figure 6. Using the Signal Builder block as the signal source, one can create the fault control signal which causes the plant model to behave abnormally. The generated outputs from the plant model are then used as the inputs to the feature model to be tested.

ADD FAULT CONTROLS TO MULTIPLE PLANT MODELS

If using the previous example again, fault control signals can also be introduced to the Signal Builder to generate fault conditions for the multiple plant models as shown in Figure 7. Note that in this case, the combinations of these fault conditions will form many more fault cases.

INTRICATE FAULT CASES CAN BE CREATED

The closed-loop interactions between ECUs add a lot of complications to implementation and testing. An ECU can not be properly designed and fully validated without the knowledge of other ECUs on the vehicle bus(es). The interaction among ECUs is often the closed-loop type so they cannot be simulated with a simple CAN tool or its equivalent. Due to these reasons, parallel programming issues (such as race conditions, synchronization, deadlocks, timeout, etc.) and vehicle network issues (such as time latency, missing messages, handshakes between ECUs, etc.) contribute to more complexity to the vehicle control system [11].

The signal builder provides the benefits of creating intricate fault cases. In the charts below, one can easily create cases where a fault happens during switched input, the fault goes away during switch being pressed and the fault happens after

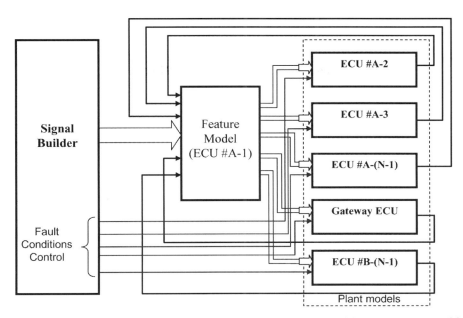

Figure 7. Multiple plant models may be placed "in-the-loop" and their fault conditions are generated by the signal builder.

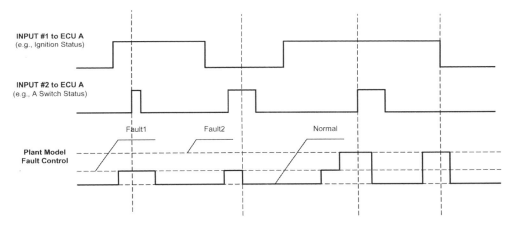

Figure 8. The Signal Builder provides an effective means to create intricate fault test cases.

the switch, which corresponds to "Fault1", "Fault2" and "Normal" cases indicated in Figure 8. Many more test cases can be created using these intricate fault cases.

REUSE FAULT INSERTED TEST CASES FOR HIL TESTING

The MATLAB tools (Simulink toolbox or MATLAB scripts) offer various means of recording the test data -from either a signal source in the *Sources of Simulink Library Browser* (e.g., the Signal Builder block) or the simulation results. All the recorded data can then be converted to a desirable format in order to reuse the test cases that have been created in

various validation phases. The diagram in Figure 9 illustrates the idea of collecting test data from the Signal Builder, the outputs from the plant models, or any inports/outports in the test harness model. These data can be stored in the simulation environment for further data disposal. Often, the data can be exported to a universal file format (e.g., the Excel or CSV format) which can then be imported to the HIL tester as introduced in [12] for hardware-in-the-loop testing or validation as shown in Figure 10. Note that all the plant models can be imported to the HIL tester using the Veristand environment [12] or an alternative tester[3] so that the hardware-in-the-loop testing is executed in an environment equivalent to the virtual test harness that has been utilized for

---

[3]A number of commercially off-the-shelf testers have the capability of importing and executing user-created MATLAB models (i.e., Simulink/Stateflow models) while sample inputs and generate outputs via tester's real-time hardware products.

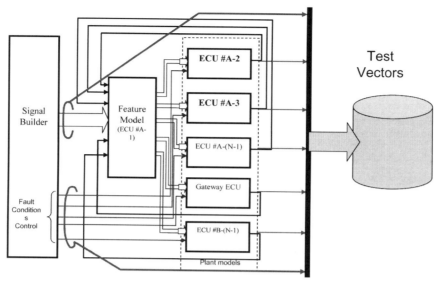

Figure 9. Test data can be recorded and/or collected from the test harness model. A data file is then generated with the data for hardware-in-the-loop testing and validation.

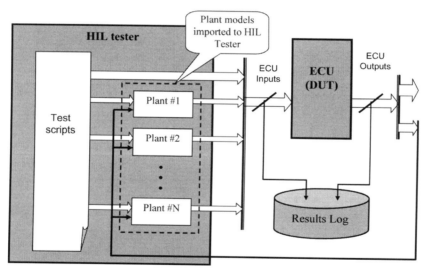

Figure 10. All the plant models can be imported to the tester so that the hardware-in-the-loop testing is executed in an environment identical to that of the test harness model.

test case development. Again, all the test data that are collected from the ECU IOs will be used to generate the results logs, which can be imported to MATLAB's environment for further analysis.

DYNAMICALLY DEPLOY TEST CASES ON A HIL TESTER

In many cases, test scenarios have dependencies on each other so that test vectors have to be deployed in a predetermined test sequence. Often, if one test case fails, it will no longer make sense to continue the rest of the test

cases. Here let us again use the *passive start* example introduced earlier. If a test case is to program a passive key (keyfob), then the failure of programming the keyfob should call off the rest of test cases that are intended to test other vehicle starting functions. In this case, certain results checking criteria need to be applied prior to kicking off the following test cases in the test sequence. Checking test results while dynamically deploy test cases "on-the-fly" (vs. static test cases) are proved to be particularly useful for testing diagnostics functions. For instance, to erase information stored in the non-volatile memory (e.g., programmed key serial data), one needs to pass certain security checks, which

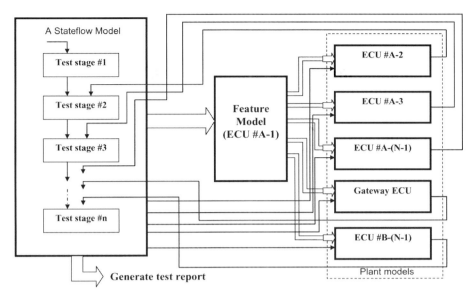

Figure 11. A Stateflow model can be used to generate test stimuli in a predetermined sequence so as to address the test sequence and dependency issues.

normally involves multiple rounds of seed generation and key submittal. As the seed data will be randomly generated and the key data will be calculated from the seed data, there is no way of using preset test vectors (i.e., static test cases) to accomplish this task.

A Stateflow model can be used to dynamically produce test vectors as well as perform results checking. Figure 11 gives an illustration of the idea. Such a Stateflow model (which can also be considered a plant model in-the-loop) is designed to address the issues summarized above which produces outputs to stimulate the ECU under test. The testing method of using such a Stateflow model also has a benefit for "fast validation", in which only a subset of the test case set are selected and built into the Stateflow model. In this way, by running a selected subset of tests, engineers can either ensure a software change not to break the already validated basic functions, or ensure a design change to be properly implemented. More importantly, this technique is particularly useful to verify the fault handling capability when fault events occur in different combinations and/or in different orders. In this case, the Stateflow model is used to deploy the fault inserted test cases in a predefined manner and to check the test results "on-the-fly".

DISCUSSION

Number of Reusable Test Cases

The test cases created in different phases are intended to be accumulated and reused for other validation phases. Because some test cases created in one validation phase may not be applicable to another validation phase, additional test cases are usually created. Such additional test cases also include those created by modifying test cases for other validation phases or for different validation cycles due to addition of new features, design change requests or design improvement. The percentage of reusable test cases for different validation phases is summarized in Figure 12 based on our production project experience.

Create Test Cases for Requirement Changes

In most cases, requirement changes are straightforward from the unit-level design standpoint but bring up a lot more complications to the overall design and system validation. This is also true even when the requirement is changed for other ECUs but not the one being worked on. Whenever such requirement changes happen, new test cases will be added to the original test set and new test run may also need to be underway. In this case, the existent test cases already used to test against aspects around the old requirements will seldom be changed or removed although test results checking criteria will accordingly need to change due to the new requirements. Rather, the old test cases will normally be retained and new test cases will be added to address the new changes. While design changes usually have the purposes of adding new features, resolving certain design issues or improving design robustness, the new changes would quite often break the already implemented/validated functions or bring up whole new issues. It is found that the strategy of *"retain and grow"* test cases is very helpful to both verify the new design and the design aspects that have been already validated. To this end, the MATLAB scripts have been utilized to maintain the test set and to add test cases required by new change requests.

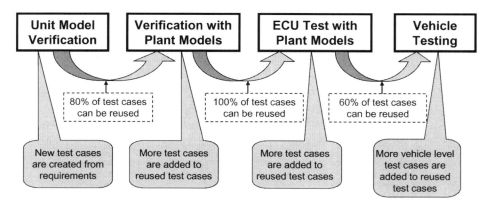

Figure 12. A summary of test cases reuse in different verification and validation stages.

New Engineering Activity- Plant Modeling

While it is beneficial to use plant models for test case development and hardware-in-the-loop testing, the development and maintenance of such plant models require extra engineering effort and budget that is usually already tight. However, it has been found that the new plant modeling activity does not notably increase the budget or affect the overall delivery timing. On the contrary, it significantly improves the engineering team's work efficiency and confidence for on-time delivery because by doing plant modeling the team usually gets much better understanding on what they try to achieve. On the other hand, when the customer sees the value in developing plant models, the customer will get more confident on the progress status. As a result, the customer will tend to reduce the number of interim releases and to streamline some project review processes, which further improves work efficiencies for both the customer and supplier. In addition, the plant modeling activity plays a crucial role in improving the product quality. Note the above statements are also true for the Stateflow model used for test vector deployment and result checking as the Stateflow model is also considered a type of plant model in this context. From the project management and planning perspective, the conventional processes such as design analysis and peer reviews are traded off for a more effective and efficient V&V process.

Design Changes and Design Variants

According to our experience, the customer will better recognize the importance and value of plant modeling when design changes and/or multiple design variants are needed. The design changes may happen any time down the road of product development and deployment while design variants can cause many unforeseen issues. Since a design change on an ECU or a different design variant may affect the overall vehicle design, there is always a need for the customer to find the potential impacts and/or issues. With the plant models and simulations, issues caused by design changes will be quickly revealed and be timely propagated to other ECU suppliers so

that all the suppliers can respond to the issues. With variants of plant models (that correspond to design variants), the same simulations are employed to identify potential issues with the design variants. After all, a supplier cannot claim its success before the customer finds the ECU successfully runs on the vehicle with other ECUs. The confidence of on-time delivery, customer satisfactory, and better supplier-customer relationship all count and are all important to stakeholders.

Seeking Customer Support for Plant Modeling

In most cases, a simplified version of a plant model will be sufficient for validating the interactive functions among the ECUs. This is because full versions of requirements specification for the plant models may not be available, and/or developing full models may not be realistic either. It has been found that OEMs normally support the effort of acquiring the domain knowledge necessary for creating plant models because efforts of verifying designs and validation products at the vehicle system level will benefit both OEMs and suppliers after all. As the customer verifies the interface behaviors between the ECU being developed and the other ECUs using the plant models simulated in-the-loop, the customer will recognize the urgency and necessity for acquiring the functional requirements necessary for developing the plant models.

SUMMARY AND CONCLUSIONS

In this paper, a model-based methodology to cope with the challenges and dilemmas encountered in developing distributed body-control applications that are implemented using multiple ECUs across vehicle networks is described. The methodology focuses on aspects of test case development, test case reuses and plant modeling for design verification and product validation. Striving to leverage executable specifications by extensively executing model-in-the-loop simulations, we develop test cases using the Signal Builder and plant models with multiple goals including requirements coverage and clarifications, design verification, hardware-in-the-loop validation, and reusing test cases throughout the entire development process.

While the role of using plant models is emphasized, we also strive to reuse plant models in both virtual test harness models and ECU hardware in-the-loop testing. The plant models may have broader implications in this context which may be behavioral models for other ECUs on the vehicle network, models for fault insertions for both the ECU under development and its surrounding ECUs, and models that can issue diagnostic commands. If the feature models in the test harness model have been the core work for implementing the ECU under development, then the development of plant models has become the focus for verifying and validating the ECU. The models imported to the ECU tester can also include models for test results checking and verification. Moreover, some test vectors that have been designed in the test harness model can be converted to Stateflow models which can act as a dynamic input vector generator to the ECU under test. Such test cases in the form of models have been proved to be a very effective means to duplicate and identify root causes for software issues in the pre-production stage. The methodology is also very useful for a large scale modeling project when multiple large models need to go into one microcontroller.

The testing method introduced in this work is not intended to replace the traditional integrated in-the-loop testing methods (such as a plywood buck testboard where all the ECUs are integrated). Rather, the test method here provides an alternative that facilitates fault insertion testing, automatic testing, and reusing test cases created in a virtual model-based testing environment. The future work may include more study on plant modeling, including methods of simplifying plant models, maintaining changes for plant models, and strategically getting OEMs involved in the process so that MATLAB models with interactive functions can be shared among suppliers.

REFERENCES

1. Amsler, K.-J. Dr., Fetzer, Joachim Dr., Lederer, D. Dr., Wernicke, M., "Virtual Design of Automotive Electronic Networks - From Function to ECU Software", Automotive Electronics I/2004, special issue of ATZ, MTZ and Automotive Engineering Partners, Vieweg Verlag (Wiesbaden, Germany), p. 2-4.

2. Navet, N., Monot, A., Bavoux, B., Simonot-Lion, F., "Multi-source and multicore automotive ECUs - OS protection mechanisms and scheduling", Invited paper at the IEEE International Symposium on Industrial Electronics (ISI 2010), Bari, Italy, July 4-7, 2010.

3. Fürst, Simon, "Challenges in the Design of Automotive Software", BMW Group, 80788 Munich, Germany, 2010 EDAA.

4. Hegde, Rajeshwari, Gurumurthy, K S, "Model Based Approach for the Integration of ECUs", Proceedings of the World Congress on Engineering 2008 Vol I, WCE 2008, July 2 - 4, 2008, London, U.K.

5. Dillaber, E., Kendrick, L., Jin, W., and Reddy, V., "Pragmatic Strategies for Adopting Model-Based Design for Embedded Applications," SAE Technical Paper 2010-01-0935, 2010, doi:10.4271/2010-01-0935.

6. SAKAI, Kunihiro, NISSATO, Yukihiro, and KANEDA, Masahiro, "Comprehensive ECU Testing Using Simulation Tools", Mitsubishi Motors, Technical Review, 2007 No. 19.

7. Mitchell, B., "Characterizing Communication Channel Deadlocks in Sequence Diagrams", IEEE Transactions on Software Engineering, 2008, 34 (3). pp. 305-320.

8. Williams, B. C., "Model-Based Programming of Intelligent Embedded Systems and Robotic Space Experts", Proceedings of IEEE, Vol. 91, No. 1, January 2003.

9. Yang, J., Bauman, J., and Beydoun, A., "An Effective Model-Based Development Process Using Simulink/ Stateflow for Automotive Body Control Electronics," SAE Technical Paper 2006-01-3501, 2006, doi: 10.4271/2006-01-3501.

10. Yang, J., Bauman, J., and Beydoun, A., "Requirement Analysis and Development using MATLAB Models," *SAE Int. J. Passeng. Cars - Electron. Elect. Syst.* 2(1):430-437, 2009, doi:10.4271/2009-01-1548.

11. Lee, Edward A.: "Embedded Software", Advances in Computers 56: 56-97 (2002).

12. Bauman, Jason G., LaZar, Darryn, "Lear Reduces Embedded Software Issues Using the NI HIL Platform", http://sine.ni.com/cs/app/doc/p/id/cs-12710 Accessed 06/14/2010.

CONTACT INFORMATION

Jinming Yang
jyang01@lear.com

Jason Bauman
jbauman@lear.com

Al Beydoun
abeydoun@lear.com

DEFINITIONS/ABBREVIATIONS

ECU
    Electronic Control Unit

BCM
    Body-Control Module

PEPS

Passive Entry and Passive Start

IO & IOs

Input(s) and Output(s)

HIL

Hardware-in-the-loop

IP & IPs

Intellectual property/properties

V&V

Verification and validation

MATLAB®

A modeling environment from The MathWorks consisting of a suite of software.

Simulink®/Stateflow®

The modeling tools in the MATLAB® tool suite.

Signal Builder

A functional block from Simulink® library browser.

Highly Scalable and Cost Effective Hardware/Software Architecture for Car Entertainment and/or Infotainment Systems

Hans A. Troemel, Jr. and Mike Burk
Panasonic Automotive Systems of America (PASA)

ABSTRACT

In the current automotive entertainment/ infotainment environment, subsystems are segmented by traditional technological familiarity and similarity. These subsystems include the radio space, the rear seat entertainment space, the infotainment space, and the satellite radio space. All of these subsystems are connected through the entertainment bus of the car. Simply speaking, the current architecture is cumbersome, unnecessary, expensive, and it is not scalable or easily reusable. The solution is integration by removing all of the redundancies and keeping only the technological necessities, while keeping hardware and software scalable.

INTRODUCTION

Currently, the automotive entertainment/ infotainment market has divided into many different modules or spaces. The radio space holds the traditional AM/FM function, streaming music media, and audio management, as well as the occasional navigation system. The rear seat entertainment space is focused on video entertainment management. The infotainment space (i.e. Vehicle Concierge Service, Bluetooth-based Hands-free, etc.) is located in varying parts of the car, but focuses primarily on the driver. Satellite radio is banished to the trunk due to non-traditional automotive technology, and size. All of these spaces must communicate through the automotive entertainment bus.

As one can imagine, there is plenty of redundancy, inefficiencies, and unnecessary latencies within the overall system. Since all of subsystems communicate over a bus, each module requires a micro-controller and the LAN electronics. All units must be able to withstand automotive power supply requirements for under-voltage, over-voltage, and load dump. In addition, all subsystems must have mechanical housings to hold them. The actual data, audio/video management and latency become more complex. Data must be filtered through many possible gateways before reaching its intended destination. Also due to EMI concerns, all audio and video transmission requires large voltage-swing analog signal or expensive optical systems. The duplication of fundamental support circuitry adds unnecessary system cost.

A possible solution is integration of most of these spaces into the radio head unit. Skeptics may say that this is physically impossible due to size, RF interference concerns and heat issues, but all of these concerns can be overcome. All of these spaces can be combined into the radio space by utilizing a highly scalable and cost effective hardware/software methodology. This concept is applicable from the mid-line system up to the very high-end system. By commonizing like hardware elements, understanding control/audio/video bus strategies, and adding the core elements for each new technology, the duplication of power supply, mechanical chassis, processors, and LAN hardware can be eliminated. Additionally, new technology integration can happen late in the development cycle with minimal impact to the current hardware. This notion is achieved through interfaces, such as USB, SD memory, Bluetooth, and 802.11. These interfaces allow for faster integration of consumer electronics and streaming media without affecting the robustness of the radio system.

This philosophy also applies to software architecture through the use of object-oriented programming. The hardware drivers are written once and HMI and Audio/Video content tasks can be based on scripting languages. Software updates and additional features can be added late in the program or after production with minimal revalidation of system functionality. Furthermore, customer approved third party software installation can bring greater feature content to the entertainment system and an additional revenue stream for the supplier. All of these ideas flow into a ubiquitous system architecture that achieves faster time to market, greater flexibility, and higher reliability, while also being extremely cost effective.

Today's automotive environment is rapidly changing. The consumer is expecting more electronic functionality to add to the overall driving experience. The customer expects luxury sedans to have a navigation system option. Vans/SUVs are also expected to have a DVD RSE (rear entertainment system) option. With this added expected complexity, the automotive industry has responded by creating modules and connecting these modules via the automotive entertainment bus. This direction only causes a more expensive solution for the consumer and slower mass-market adoption of new technology.

A preferred solution is up-integration into the head unit. All of these spaces can unite into the radio space by uses a highly scalable and cost effective hardware/software methodology. Throughout the rest of this paper, we will explore

- The general state of affairs of the automotive entertainment/ infotainment sector
- The commonalities of the different spaces
- The integration of the overall system
- The high level hardware aspects to this system concept
- The high level software aspects to this system concept
- A product that demonstrates these ideas and concepts

CURRENT AUTOMOTIVE ENTERTAINMENT/ INFOTAINMENT SITUATION

The Traditional Radio Space – Traditionally, the radio head unit has been the focal point for the control of the automotive entertainment experience. This space is prime real estate due to its location within the vehicle. In past years, the automotive industry only offered one form of entertainment: Audio. The audio sources came from the air via AM/FM radio stations, or from media via CD, cassette tape, or 8-track tapes. The radio is responsible for proper source selection and amplification to the speakers of the car or to headphones for the rear seat. These functions are controlled through a HMI (Human-Machine Interface) that is easy to navigate and understand, while minimizing driver distraction.

Within this traditional concept, the radio has very little information to present the end-user. In reality with the addition of RBDS, MP3 navigators, and SDARS/HD Radio information, the radio has many new HMI challenges.

Within recent years, the navigation radio has merged into this space. This market is forcing higher integration into the radio space. Navigation is the first market segment to integrate an information-based source into the radio space. This level of complexity requires advanced programming techniques. Since navigation

requires more HMI intervention and video processing, higher-end processors are a necessity for applications such as voice recognition, map database manipulation, dynamic route calculation, acoustic-echo cancellation, and audio arbitration. From the software side, this progression allows for more modern techniques to be employed, such as the separation of driver code and HMI code under the management of a scheduler, or as the complexity level raises, an operating system.

The RSE (Rear-Seat Entertainment) Space – The RSE space is a newcomer to the automotive entertainment segment. The industry ignored the idea of video in the vehicle long ago due to driver distraction. In contrast, the consumer's voice seemed to disagree as evidenced by the overall popularity of this new market segment. As long as it is in the rear seat (or keeps the kids happy), this market has enhanced the long trip driving experience and is one of the fastest growing automotive entertainment market segments.

Currently, these systems are fairly stand-alone. The DVD mechanism is co-located with the central monitor or remotely located from the radio. The audio is delivered through IR to the headphones or delivered to the speakers via the head unit/amplifier system. The HMI uses a remote control that also provides some limited radio controls. Allowing control to the viewer is a nice option, but the driver should always have central control of his car. The monitor and IR transceiver must be near (and visible) to the rear passengers, but all other items can be co-located or integrated with the radio. This also allows easy integration of parental control features. Vehicle manufacturers have noticed these issues, and a new trend has emerged where the DVD mechanism will soon join the radio space.

The Satellite Radio Space – The SDARS industry is a real newcomer to the automotive entertainment market. Currently, this subscription-based service primarily offers over 100 channels of CD-like quality audio with a continental national coverage area. Both XM and Sirius are working hard to bring more than just audio to their subscriber-base with demonstrations of on-demand information services and compressed video services.

Currently, the SDARS receivers have been integrated into the vehicle body itself. Due to the lack of integration of the first chipsets, the first generation receivers were fairly bulky, making it very difficult to integrate into the radio head unit.

The control of these units is through the vehicle bus, and the audio is delivered to the auxiliary audio input of the head unit. The title and artist information is transmitted through the vehicle bus to be displayed on the radio's display. Recently, the newest versions of the receiver are small enough to make integration possible.

The Infotainment (Telematics) Space – This has recently been the golden horizon of automotive electronics, although the lack of a ubiquitous continental wireless data network has tempered the rollout of this concept. This market space is built on two technologies: the cell phone and GPS. The cell phone offers a two-way wireless voice and data communication link to a service provider. The GPS receiver offers location and speed information. The location information is transmitted through the cell phone data link and is utilized mainly for 911 emergency location, remote server navigation information, and concierge services. One approach to this service is Onstar. Onstar gives services of Safety and Security (Emergency Helpline, Automatic 911 services), Concierge services (Directions, Reservations, and Door Un-lock) through their call-in service center. The consumer gets to talk with a live person to obtain the information and can even use the embedded cell phone for private use for additional monthly fee. The integration of phone technology, GPS, and speech recognition offers the possibility of enhanced automated concierge services. This architecture can improve the return on investment of concierge services.

A highly desired consumer feature is to allow voice communication with minimal driver distraction. Bluetooth is a short range networking wireless communication system that offers features such as Hands-free cell phone communication, synchronization with handheld devices (PDAs, Laptops), networking capability, and limited streaming media applications. Bluetooth can be imbedded into the radio head unit and act as the master of the pico-network. Even though this technology does not replace the service provider, it does offer greater opportunity to enhance the current call center feature and offer greater automation.

Telematics is based on telecom rather than traditional automotive technology; thus these products were developed as stand-alone modules. Most of these modules utilize a separate control cluster. The microphone goes directly to the stand-alone module for voice recognition and voice processing. The cell phone and microphone audio is delivered to the radio head unit through an auxiliary input that is overlaid onto the appropriate speakers. Even though the suppliers usually operate from a different market segments, the technology behind the products is very similar to voice controlled navigation systems without the embedded cell phone.

The Automotive Bus – The glue that holds this overly complex system together is the automotive bus. As you can imagine it has quite a bit of traffic to manage. Buses such as CAN are becoming more prominent. Unfortunately, the CAN bus is not fast enough for most streaming media (Audio & Video) applications. For these applications optical buses such as MOST are employed. The primary advantage of optical (vs. electrical) buses is their superior EMI performance. If the cost impact of the optical bus outweighs its advantages, the wide-band streaming media connection can be provided via analog point to point structures. Either method makes adding high feature content and

functionality to the automobile a complex, expensive effort.

INTEGRATION FROM A SYSTEM PERSPECTIVE – The current system dilemma is illustrated in Figure 1. An answer is integration of most of these spaces into one common space: the radio head unit. This idea makes sense due to the prime real estate and ease of command and control from the driver's perspective. Through this integration, system redundancies are eliminated, the robustness of the overall system becomes a Tier 1 responsibility, and unnecessary costs are eliminated from the overall system. The real question is "How can a Tier 1 supplier support all of these markets without re-inventing the wheel every time?" First, let's take a look at this problem from a general system perspective.

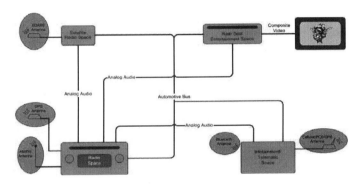

Figure 1. Current System Architecture

System Redundancies – First of all, let's be fiscally responsible by exploring the opportunities to save money by eliminating redundancy from the system. Each of these subsections does contain the following elements: LAN circuitry, a micro-controller, automotive-grade power supply, streaming media connection, and mechanical chassis/components. If this entire system is integrated into the traditional head unit space, all of these redundancies can be removed. To illustrate, for each additional micro-controller and LAN interface, additional software, hardware, systems, and test engineering resources (possibly from different companies) must work together to ensure that point/function X is communicating nicely to point/function Y. In an integrated design, many of these interfaces simply go away, or are reduced to communication between hardware and/or software modules residing within the same product!

In addition, each of these separate modules would need to be validated to the robust environmental standards of the automotive industry. Also, each streaming media connection eliminated would save the audio DAC, buffer/amplifier drivers, the associated differential receiver, and re-digitizing ADC to ensure improved signal quality and increased dynamic range. Although the resulting head unit cost may be somewhat higher than the traditional HU design, the overall system cost will be lower than the total cost of all components and vehicle overhead (wiring, etc.) for the traditional design with less functionality.

Technology Commonalities & Differences – Another system aspect to consider is the technology commonalities of the overall system. In other words, many of the modules share similar technologies or enabling technologies. Some modules share similar algorithm features. For example, both the infotainment space and the (navigation) radio space require voice recognition. This algorithm requirement drives each of these modules to have an interface to a microphone, an acoustic echo-cancellation and a noise-cancellation algorithm, and voice prompts. A bigger technology similarity shared between all spaces is the audio DSP algorithms. The RSE, infotainment, satellite radio spaces perform audio DSP processing on their respective sources, while the radio space is the final audio processing stage before interfacing to the speakers.

The differences are more isolated for the specialized analog circuitry. Even though in concept RF front-ends for GPS, AM/FM, SDARS, Bluetooth, 802.11, and Cell phones are very similar, the differences in the detailed communication system requirements tend to favor traditional narrowband RF circuit design techniques.

The video of the RSE space is also a unique feature. Since the monitor needs to be close to the viewer, these signals need to be delivered as either optical or analog signals based upon EMI and cost concerns. Furthermore, video is a broadband signal and requires a different set of circuit design techniques.

Integrated System Concept: Feature Progression & Migration – All of these redundancies, similarities, and differences must be taken into consideration for integration into the radio space. The proposed concept is shown in Figure 2. This figure shows all the media sources, RF sections, and audio/ speech processing integrated into the radio space. Only the antenna system, video monitor, and remote control are external to the radio head unit. Added features and entertainment media may be added through common consumer electronics ports, such as Bluetooth, 802.11, USB, SD memory, 1394 ports, and a front panel stereo audio jack.

This system architecture starts with the basic element of the automotive radio: the DIF (Digitized IF) AM/FM tuner and audio DSP processing. These two elements are the foundation of the system. Since all streaming media source material is digital, eliminating any intermediate (and unnecessary) analog to digital conversions will maintain superior audio and video quality, ease migration between feature complexities, and take advantage of the Moore's Law of cost reductions. Additionally as speech algorithms, SDARS/HD radio processing, HMI complexity, and networking needs increase, the processing components are scaled or distributed based on simultaneous processing requirements. Take for example SDARS processing needs to be running in tandem with other processes due to buffering requirements and processing complexity for acceptable source switching requirements. On the other hand, processor bandwidth can be reused between mutually exclusive product features.

This system architecture is scalable based on product feature set. As seen in figures 3, 4, 5, 6, and 7, the same system concepts apply from the very basic ETR system to the most complex entertainment/infotainment systems. These figures demonstrate the powerful notion that only the core technology elements or differences are added as the feature content grows. This idea leads to greater standardization of hardware and software drivers.

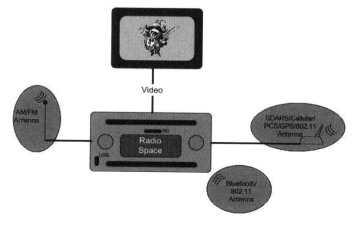

Figure 2. Proposed System Architecture

ETR System
- *Core DIF Radio*
- *Base Audio Package*

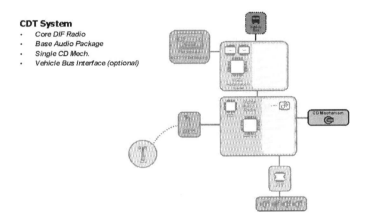

Figure 3. Basic ETR Platform

CDT System
- *Core DIF Radio*
- *Base Audio Package*
- *Single CD Mech.*
- *Vehicle Bus Interface (optional)*

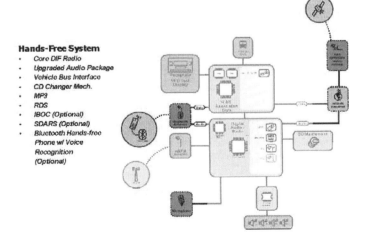

Figure 4. CDT/ICDX Platform

Hands-Free System
- *Core DIF Radio*
- *Upgraded Audio Package*
- *Vehicle Bus Interface*
- *CD Changer Mech.*
- *MP3*
- *RDS*
- *IBOC (Optional)*
- *SDARS (Optional)*
- *Bluetooth Hands-free Phone w/ Voice Recognition (Optional)*

Figure 5. Hands-Free Platform

DVDX System
- *Core DIF Radio*
- *Premium Audio Package*
- *Vehicle Bus Interface*
- *DVD Changer Mech.*
- *DVD-A*
- *DVD-V Output*
- *MP3*
- *RDS*
- *Audio EQ*
- *IBOC (Optional)*
- *SDARS (Optional)*
- *Bluetooth Hands-free Phone w/ Voice Recognition (Optional)*

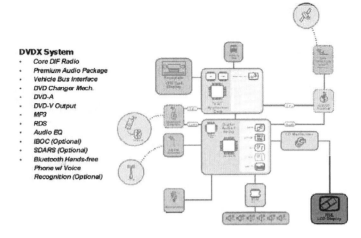

Figure 6. DVDX Platform

Infotainment System
- *Core DIF Radio*
- *Premium Audio Package*
- *Vehicle Bus Interface*
- *Single CD Mech.*
- *DVD Changer Mech. (Optional)*
- *DVD-A*
- *Media Storage*
- *DVD-V Output*
- *MP3*
- *RDS*
- *Audio EQ*
- *iBOC*
- *SDARS (Optional)*
- *Bluetooth Hands-free Phone w/ Voice Recognition*
- *Navigation w/ GPS*
- *Phone Module*
- *Remote Connectivity*
- *Security Features*
- *Text-to-Speech Feedback*

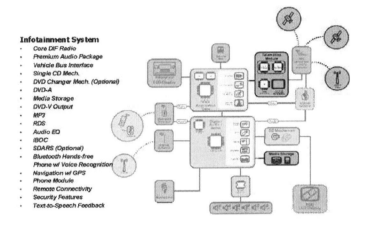

Figure 7. Infotainment Platform

HARDWARE INTEGRATION CONCEPTS – To make this system architecture a reality, hardware specifics such as identification of key hardware elements, distributive and scaling processors, buses, plug & play hardware blocks, and connection portals must be further explored.

<u>Key Hardware Elements</u> – From the system perspective, this new system is highly integrated and scalable. In order to keep this system perception, only development of the key technologies are developed at a chipset level. Furthermore, scaling and/or distributive processing techniques must be assessed for all processor elements.

DIF AM/FM Tuner with Audio DSP Processing – As mentioned earlier within this paper, one of the keystones of this notion is the DIF AM/FM Tuner/ Audio DSP. Even though the AM/FM tuner itself really fits into the RF peripherals category, AM/FM music is the most basic element of an automotive entertainment system.

Within the last few years, most of the digital controlled AM/FM tuners are composite signal (MPX) based architectures. This architecture is shown in Figure 8. Within this architecture, the AM/FM RF signal is down converted to undemodulated baseband signals, then these baseband signals are digitized. The DSP performs noise blanking functions, demodulation, and blend/ roll-off parameter manipulation. The dynamic IF filtering is controlled by the DSP, but is still accomplished utilizing analog techniques within the tuner. Additionally, over half of the functionality of this DSP is devoted to audio processing, source selection, and sample rate conversion.

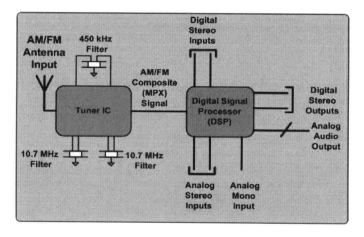

Figure 8. MPX AM/FM Tuner Architecture

The DIF (digitized IF) architecture fundamentally moves the digitization of the AM/FM signal to the IF (intermediate frequency). This architecture is shown in Figure 9. This change moves the dynamic IF filtering process to the DSP. Moreover, the DIF architecture reduces the number of components in the AM/FM tuner section, and allows Moore's Law of digital circuitry greater opportunity in reducing overall costs. More audio connections can remain in the digital domain allowing better separation and signal-to-noise. Furthermore, these connections are easier to implement, less susceptible to noise, overall a lower implementation cost. Features such as audio arbitration, car fixed equalization, user equalization, dynamic loudness, compressors, and expanders are easily reconfigurable for each vehicle manufacturers needs.

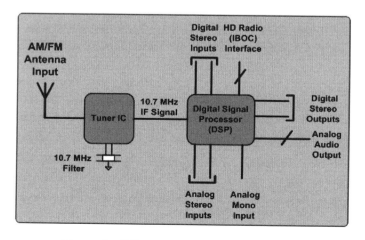

Figure 9. DIF AM/FM Tuner Architecture

Moreover, the DIF architecture adds improved reception quality and one more future feature: HD radio. HD radio is becoming a reality for the AM/FM broadcasting world. Since HD radio is an In-Band On-Channel (IBOC) technology, the RF signals co-exist with the analog signal and occupy the adjacent channels. This changes the RF landscape and forces the receiver manufacturer to require better IF filtering techniques to maintain the same reception quality. Since these IF filters are based upon FIR algorithms, no distortion is added to the AM/FM signal with narrower IF bandwidths. Additionally, these digital IF filters almost double the adjacent channel interference rejection ratio and increase the quality of the AM/FM signal. DSP algorithms provide better tracking of the carrier frequency allowing greater signal-to-noise ratios to be achieved. Practical four quadrant switch diversity and phase diversity algorithms can be implemented to increase reception quality.

RF Peripherals – Due to the mobile nature of the automobile, many entertainment and information sources must be received through RF sources. Narrowband RF receivers offer optimum RF performance for necessary OEM quality. For this paper, the RF peripherals are GPS, Cell phone, Bluetooth, a SDARS receiver, 802.11, and AM/FM radio. All of these RF peripherals are highly reliant on the processor & DSP subsections, since the actual demodulation of the signal is no longer an RF responsibility.

Since most of the signals are external to the vehicle, these receivers interface with an externally mounted antenna. The only possible exceptions are the networking RF peripherals: Bluetooth and 802.11. Both of these technologies are low power and short range networking technologies for in-vehicle applications. In these cases, the antenna will most likely be embedded into the faceplate of the radio. Remember, each RF section works with microvolt to sub microvolt signals. This drives the overall board design; EMI requirements, processor and switching power supply clocks.

Power Supply – With all of these different features and functions, current consumption is increasing, power supply voltages are decreasing, and heat dissipation is on the rise. The two main contributors are the actual power supply section and the audio power amplifier. Traditionally, multi-voltage linear regulators offer a greater than 10W power supply for a reasonable cost, but these devices must be attached to a heat sink. Boost switching power supplies are necessary for the high voltages required by VFD displays. The traditional designs for these switching supplies have enough energy to disturb most RF sections and spray enough radiation to not pass most domestic OEM requirements.

The answer to this dilemma is changing the power supply design practice. Utilization of buck switching power supplies for major current loads and/or low voltage large current loads reduces heat dissipation. Radiation can be controlled through good layout techniques, good power and grounding strategies, using toroidal inductors within the main switching regulators, and distribution of the switching energy to multiple switching regulators. Additionally, another big key is management of the switching frequency. All switching frequencies need to be synchronized and changeable. The phase of the switching frequency is important as well. These concepts consist of RF frequency dodging or switching frequency dithering. Furthermore to help balance the switching currents, the total AC current of the total power supply should approach zero through the use of the phase of the switching frequency. Since these power supplies offer efficiencies greater than 80%, this method offers even greater flexibility with different load configurations that are necessary for this architecture.

The other main contributor to heat dissipation is the audio power amplifier. Almost one third of the total power consumption is dissipated in heat. Class-D amplifiers offer a nice solution to this problem. Unless the vehicle manufacturers demand this technology, its adoption will be slow due to cost, real estate size, and radiation issues.

Media Drives – Bringing personal media into the vehicle is not a new concept. The 8-track cartridge was popular in the seventies. The cassette tape started in the eighties and is finally fading out of the automotive entertainment system. Currently, single CD and in-dash CD changers, and DVD drives, dominate the media market. These drives also have the capability of compressed audio, like MP3. Each of these drives has their own power supply, command/control, and streaming media requirements that must be taken into consideration for the overall system. These devices are stand-alone or treated like an RF peripheral. In other words, strip out the traditional DSP and D/A conversion from the mechanism, and only have the bare mechanical component, the optical interface, and motor control electronics, while most of the signal processing is integrated into the main board electronics in the digital domain. In the future, these drives will change once again as hard drives are integrated into the radio.

Processor & DSPs – For this proposed system architecture, processors and DSPs need to stand-alone. This subsection has all of the basic principals - Some things are common, some things are scaled, and interface connections are digital. One of the main cost savers and system simplifiers is the integration of all of these module processors. Naturally, all of the command/control, application data, and streaming media interfaces have a propensity to be digital.

By far, the chief concept is scaling. Processors and DSPs are sized to complexity of display, HMI, and feature applications. From the feature matrix and HMI definition, the parallel processing requirements are calculated. Even though a single very fast processor is a nice idea, this idea has a tendency to not be practical; therefore scaling versus distributive processing is inevitability. Part of the balance lies in the origin of the

algorithm. In order to sell more components, some integrated circuit suppliers have developed algorithms to run only on their chips. The acceptance of this perception places these processes in a distributive manner. Furthermore, the OEM's selection of algorithms such as voice recognition, acoustic, and noise cancellation algorithms drives a balance between DSP and micro-control processors. As a result, the radio supplier needs both hardware options to suit the specific needs of the OEM. Moreover, this processing balance also extends across product families. By sticking with processor families, like ARM, the same code set and capabilities are shared between processor sizes. For example, the ARM-7 family is a good-sized processor for even low-end radio products, while the ARM-9 family offers the power for high-end applications. In this particular case, the ARM family is also utilized in the consumer electronics market allowing the automotive market to leverage the cost benefits.

Hardware Buses – One of the last major hardware considerations to make this highly scalable architecture a reality is consideration of interconnect. Optimization of command/control buses, application data transfer, and streaming media is achieved through standardized bus selection. Of course, the final selection is limited to the available options of each of the key element subsystems.

Command & Control – In the case of strictly command/control bus, the data rates are low. An application example is CD mechanism control. Due to the low data rates, this bus can be time-multiplexed with other peripheral subsystems. As always, bi-directional communication and minimization of bus wires is preferred. An example of this view is the IIC bus developed by Philips Semiconductor. This two-wire bus has only a clock and data lines and can support data rates of 400 kbps. Since all slaves have an address, this bus is easily time multiplexed. Also, this bus has a multi-master mode for communications or data transfer. Plus, an ACK signal exists for each packet for hardware handshaking. For further information, please visit Philips Semiconductors website at www.philips.com.

Application Data - In the case of application data and command/control information, the data rates are medium to high, and two-way communication is a necessity. An application example is Bluetooth. These buses need to be dedicated buses. As always, minimization of bus wires is preferred, but data rates argue for synchronous bus structures. An example of this view is UART bus or SPI bus. The UART bus can be synchronous or asynchronous. The asynchronous bus uses only two wires, while the synchronous bus is a four-wire bus. The SPI bus is also a synchronous four-wire bus. These buses can support data rates of greater than 1 Mbps. Handshaking is accomplished through software and/or hardware (for synchronous structures). For further information on SPI bus, please visit Motorola Semiconductors website at www.motorola.com.

Streaming Media – As mentioned earlier, streaming media is defined as audio or video streams. These streams are traditionally analog. This concept limits the quality due to the high susceptibility to noise and should be limited only to low-cost module interconnections. But with the proposed architecture, these buses are internal to the radio and can remain digital for high quality and greater flexibility. For example, stereo audio can be transferred using IIS or SPDIF. Both buses support 24-bit stereo information and are uni-directional. The SPDIF is a single-wire bus, while IIS is a 3 to 4 wire bus. For two-way voice communication, a PCM bus is a four-wire bus commonly used within the cellular industry.

Plug & Play Hardware Blocks – Just like any complex system, key technologies must somewhat stand-alone for development. Each of these key technologies has it's own set of concerns to optimally work within the overall system. In order for smooth integration, the key technologies will have standardized digital interfaces for all interconnections. This action will allow for "snap together" or hardware plug & play feel for the overall product design utilizing proven key technology hardware blocks.

As many of the Tier 1 manufacturers recognize, some key technology hardware is extremely layout sensitive. To prevent constant rework, some ECAD systems offer a key feature that binds a layout sensitive schematics and it's respective artwork. A specific example of this type of tool is the Reuse Module feature of the Mentor Graphics system. This tool allows for the AM/FM tuner schematic and layout to be treated as a frozen hardware component. The tuner section is placed as a whole onto the circuit board. This capability allow for greater repeatability from product to product. Also, different layout versions of the same schematic subsection can be created based on differences in interface connectors or single reflow versus dual reflow requirements. In any event, this key technology can be verified as a stand-alone entity and easily integrated/shared within many product designs.

Expandability Connection Ports – For year's automotive corporations not only tried to find new ways to sell vehicles, but also expand into on-demand services. Theses goals can be accomplished through consumer electronics portals, such as Bluetooth, 802.11, USB, and SD memory card. All of these technologies tied directly to the processor elements and their processor usage must be taken into account. Bluetooth technology is fully integrated into the faceplate of the head unit. The USB and SD connectors are easily accessible to the consumer.

These portals can be utilized for new media information, for example MP3, WMA, and/or MP4 video streams. Software updates and additional features can be added after production with minimal revalidation of system functionality. Besides, OEM approved third party software installation can bring greater feature content to the radio and an additional revenue stream for the supplier. Also, these technologies continually migrate and change over the life of the vehicle that continually improves the value of your product.

SOFTWARE CONCEPTS

Componentized Software – One of the primary goals of modern software techniques is to produce reusable code. Unfortunately, this target most often is not realized. Software components represent pieces of code that can be retained to be used on subsequent projects as pre-built, pre-tested functional blocks. In a larger environment, such as a PC, these components might be pre-built binary components using CORBA or COM. In the embedded environment, software components should be source code components to be compiled into future products for the same advantage, but consuming less space.

One of the primary goals of componentizing software is to produce a software component to go along with a hardware component wherever that hardware component may be used. For example, the software for controlling a CD mechanism should be able to follow that CD mechanism into multiple products or even multiple product lines with no modification. We don't build a new CD mechanism every time we build a new head unit, why should we build new software to run it?

Compartmentalized Functionality – To accomplish this methodology, we must compartmentalize functionality. This means that functionality must be separated into independent components with rigid demarcation lines between components, and well-defined interfaces between components. This concept is not just limited to the software components that correspond to hardware components, but to logical components that represent an associated set of functionality.

The strong demarcation lines between components:

- Simplifies development and unit testing.
- Improves maintainability.
- Improves extensibility.
- Improves reliability.

Add to this the adoption of well-defined interfaces early in the development process for more refined benefits:

- Allows parallel development on either side of the interfaces and demarcation lines to help beat the "Mythical Man Month". Different people, different teams or different contractors can be working in parallel with full knowledge of how the component should work and what test criteria they must meet.
- Proper interfaces allow for mixed languages. Some components can be written in C, some in assembler, some in C++ (EC++).

- The interfaces could plug into a scripting engine or virtual machine so that the HMI could be done using a scripting language or Java.
- The interfaces could be exposed in such a way as to facilitate use by auto-generated code.

<u>Lower Engineering Cost</u> – This methodology reduces engineering costs.

- Less debugging due to better design, and code re-use
- Less testing due to re-using pre-tested components
- Higher reliability as a result of the better designs and re-use of critical components
- Higher quality due to the better design and code re-use

<u>Object Oriented</u> –Keep in mind that Object Orientation is a concept, not a language requirement. You can do Object Oriented Design (OOD) then code in any language you choose, even assembler. Object Oriented is the way we as humans naturally think. We think of things that have characteristics and are able to perform certain tasks. That's exactly what Object Oriented programming is. In comparison, hardware by nature is Object Oriented. Chances are an embedded developer who has a hardware background, already programs in this manner without realizing it.

Using Object Oriented methodologies allows for better separation of the user model and the hardware model. This separation is something that is not a critical issue in just a radio, but as functionality increases and diversifies, lack of separation will become more of a problem. A manifestation of this problem is the tendency for developers to write the program to match the hardware model. The developer must keep in mind that the programmer should learn to think like the consumer rather than requiring the consumer to think like a programmer. Using an Object Oriented language, such as Embedded C++, and an RTOS (Real-time Operating System) will aid in componentizing and compartmentalizing the software. Avoidance of some of the more esoteric functionality of C++ can produce code that is as tight and fast as standard C code. But the main thing the developer needs to learn is Object Oriented thinking, and then the language and RTOS will make the development easier. An example of this software architecture is shown in Figure 10.

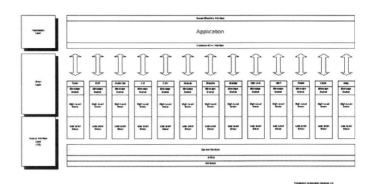

Figure 10. Example Software Architecture

AN EXAMPLE – Multifunction Infotainment Head Unit

Figure 11. Multifunction Infotainment Head Unit

A prototype product based on the architectural principles described in this paper has been realized. The faceplate is shown in Figure 11. The goal of this exercise was to prove, in physical form, that extremely high hardware functionality and very complex feature functionality could exist within the bounds of a standard two DIN chassis. The product is a mixture of audio/video entertainment functions, connected information, telematic, and advanced reconfigurable HMI techniques. This product has been demonstrated and well received at various trade events as well as in direct meetings with vehicle manufacturers.

Core Physical Elements –

The following subsystems are contained:

- DIF AM/FM radio with HD Radio capability and antenna diversity
- 6 Disc CD/CDR/DV/DVDR / MP3 changer
- Integrated XM receiver
- GPS
- Bluetooth transceiver
- Audio/Video signal processing hardware
- Application processing and memory
- DSP processing
- Tri-color VFD Graphics display
- 1.5" TFT display
- Vehicle Bus Interface
- Power Management

Feature Set - User feature applications were written to take full advantage of the physical elements. Many features are a combination of internal elements as well as a combination of on-board and off-board functions.

These features were defined in consumer research studies, modeled in a virtual prototyping environment, and subsequently launched as a physical prototype.

The following features are enabled:

- AM/FM radio with metadata capability from either HD radio or server based off board data
  - Ability to timestamp broadcast media and conduct B2B functions with service provider
- MP 3 playback with metadata
- CD/CDR with metadata
- XM radio
- Visual display of logos, cover art, etc. associated with any of the above content either locally stored or downloaded from an external service provider such as Gracenote CDDB®.
- DVD video sourcing for an externally packaged NTSC color display with full content navigation
- Bluetooth Hands Free Phone
  - Speech Recognition
  - Text to Speech
  - Data Exchange
  - Echo Noise Cancellation
  - On board message attendant
- Secondary Embedded Bluetooth Phone
  - Connection to remote data service
  - Concierge functions
  - External access to vehicle function
  - Security Functions
- Turn By Turn Navigation
  - Parsed data locally stored from server based data source
  - GPS Servo and Dead Reckoning
  - Voice Prompts
  - Embedded secondary navigation routing (navigation within navigation)
  - Navigate to direct input address
  - Navigate to phone number with reverse lookup
  - Navigate to point of interest
  - Ability to preset and label destinations for one button navigation
  - Color Logo Display for POI elements
  - Voice or manual input
- Ability to receive text based information such as news weather or traffic and playback using free form text to speech. Also color logo data may be downloaded to associate the information item.
- Security Features – The vehicle can use this product as an automated security and concierge attendant. A combination of speech recognition, text to speech, GPS, and external connectivity are used to achieve the following features:
  - Utilizing the GPS, Geo Fencing automatically calls out if GPS or speed limits are violated
    - Valet Lockout
    - Parental Lockout

- o Vehicle Tracking
- o Vehicle Status
- o Voice call to driver
- o Remote door unlocking

Obviously, this functionality is on or above the levels achieved with traditionally executed systems comprised of a multiplicity of bus-connected modules

HMI Strategies – A second aspect of this project was to study HMI strategies for highly complex functionality. The execution selected embodies a control methodology that is very similar between functions. A combination of flat menu structure and redundant interfaces make this device easy to learn, and easy to operate.

Visual Interface – A combination of a VFD multicolor display associated with a small TFT color display was selected. This combination allows high visual impact, and the ability to observe status quickly through the use of icons. Additionally this combination enables free association from surrounding soft keys that creates similar functionality to touch screen devices at a fraction of the cost.

Tactile interface – A jog shuttle with center push button resides on the side of the faceplate closest to the driver. Groups of functions for most features can be accessed by this one control

Direct Access Buttons – Paramount to this HMI execution is the ability to move freely and directly between features. This methodology eliminates deep menu structures. Sub group functions are one level down under each functional group. The last used feature remembers the last accessed function from a particular group. This functionality eliminates multiple button presses.

Multilevel Interface – The device utilized a combination of visual, audible, and tactile interface techniques to minimize driver interaction and distraction. For the most part, voice interaction is secondarily backed up by tactile input.

Customization and Rescripting – A great deal of care was taken to keep the HMI mechanical and logical strategy neutral. As a result, it is possible to add and remove features without affecting the tactile layout. The possibility also exists to take the same hardware and script it for multiple unique vehicle applications. The modular nature of the firmware allows adjustments to application functionality at any time during or after development.

Yet another highly powerful aspect of this development methodology is to add features and functionality after the point of sale. A powerful example of this technique actually occurred during the development of this product. Initially this product definition did not include DVD rear seat video functionality although the deck mechanism supported video feed. One morning a sales person frantically approached the development team with a request to add video functionality to the system to support an impending customer visit. Even though video was never part of the definition, video with complete menu functionality was added in less than 3 hours with no change to the mechanics of the faceplate!

Cut the development time! - The power of the design methodology described in this paper is a key enabler to a vastly more efficient product development. When associated with model based product development; these techniques can realize a 50-70% reduction in the development timeline. As a validation, the product described here was virtually modeled. Once the functionality was agreed upon, the physical execution of a fully functional prototype was complete and demonstrated in 120 days.

CONCLUSION

The current state of automotive entertainment and infotainment systems are compartmentalized, redundant, overly complex, and expensive. There are many reasons for this current state, but integration into the radio space is the means to solve many of these issues. The key concept of this new architecture has simple fundamental ideas: some things are common, some things are scaled, and connections are digital. These notions also extend to the software architecture through the use of object-oriented methodologies. The culmination of these approaches impact the radio design to offer greater reliability, increased flexibility, faster time-to-market, and lower overall costs through standardization within a high scalable hardware and software architecture that spans through many market product levels.

ACKNOWLEDGMENTS

Cary Wilson – VP of Engineering at Panasonic Automotive System of America

Mike Burk – General Manager of Advanced Development at Panasonic Automotive System of America

CONTACT

Hans A. Troemel, Jr.
Senior RF Engineer
PASA Advanced Development Group
troemelh@panasonic-mcusa.com

Steve Mobley
Senior Software Architect
PASA Advanced Development Group
mobleys@panasonic-mcusa.com

Mike Burk
General Manager
PASA Advanced Development Group
burkm@panasonic-mcusa.com

Analysis of Interfaces and Interface Management of Automobile Systems

Ralf Fritzsche

Ford Motor Company
John-Andrews-Development Center

ABSTRACT

Automobile systems are often linked with a high amount of interfaces in a vehicle. An essential assumption for a trouble-free application is a clearly defined information-flow over the life time period of a product. If any information is not included or not communicated to the responsible technical departments the consequence may be an unintended visit to a repair shop or perhaps even a vehicle recall. To avoid such uncomfortable situations Ford Motor Company applies and recommends a systematic approach to analyze all affected interfaces for successful system integration. The primary task of this article is to present all necessary steps and tools-in-theory. For demonstration an ideal example is attached which can be used as a practical guideline.

INTRODUCTION

By reflecting back on the development in the automotive industry over the last 15 years a lot of technical innovations of the past are now standard. Active safety devices such as airbags, ABS and ESP or comfort systems like climate control which were part of the luxury class are now integrated into the standard equipment of small cars. This development has only been possible with the use of electronic controllers and intelligent networks which control the data highway via a CAN-Bus. As a logical consequence, engineers and technicians have to detect, agree and integrate a high number of Interfaces. The automobile business trend is progressive and with every face-lift new systems or vehicle attributes with comfort electronic, driver assistance, telematics and infotainment are being offered by the car manufacturers. The objective of this technical trend is to provide the driver a higher amount of safety, comfort and fuel consumption reduction. All of these systems can react or warn the driver autonomously or partly autonomously about the behavior of powertrain operation, controllers and signaling devices. This will ensure that future vehicles are able to detect safety critical traffic situations prior to any driver reaction and can create counteractive measures with the aim to protect passengers or other traffic participants. This can be accomplished by a number of sensors, which measure constantly the actual driving conditions and inform the necessary controller.

The following is a list where the methodology can be applied:

Electronic Comfort Systems:
- Climate Control,
- Steering Wheel,
- Lighting (internal and external),
- Seat Adjustment,
- Climatic Seats,
- Tailgate or Lift Gate Actuation,
- Access Control to the car.

Electronic Safety Systems:
- Airbag Controller with or without over-roll sensing,
- Pedestrian Protection,
- Driver Detection,
- Linkage between "active" and "passive" Vehicle Safety.

Electronic Driver Assistance Systems:
- Adaptive Cruise Control,
- Blind Spot Recognition,
- Lane Change Assistant,
- Lane Departure Warning,
- Lane Keeping,
- Stop & Go,
- Active Chassis (e.g. damper control system)
- Electronic Steering Aids,
- Tire Pressure Monitoring (active and passive),
- Traction Control System (ABS, ESP, All Wheel Drive),
- Automatic Emergency Braking,
- Brake Assistant.

With the increase of functional interfaces there is a potential risk of new failure sources and the introduction of new unknown failure modes. A common root cause for this complex situation is due to the transsectoral working environment between different departments. If there is no early exchange of interface information, awkward field actions might be the consequence. The number of vehicle recalls worldwide has increased significantly during the last couple of years. A cleverly adjusted Interface management control system can resolve this cost intensive trend.

MAIN SECTION

DEFINITION OF AN INTERFACE

What is a good definition of an Interface under technical aspects especially for the automotive industry?

PAHL and BEITZ [4] are considering the interaction between energy, material and signals for a closed system. A large amount of calibrations have to be adjusted so that an active Interaction of systems or individual parts is possible. These might affect in respect to active or passive elements. In this context there are "functional interfaces" or "data interfaces". With every transfer of the above mentioned values these must be considered within their quantity and quality to define clear criteria to specify the task.

FORD [1] adopts the good approach of PAHL/BEITZ and adds a physical toughing condition. Furthermore the expression "signal" will be replaced by the more generic term "information". Within this refinement an interface can be fully described in a four quadrant illustration (Fig.1).

Physical Toughing

Any mechanical combination of two parts is described as physical toughing. Some practical examples and some ideas out of schoolbooks [5] are summarized in the following list:

Ensure a mechanical connection:
- Force-Fit connection: screw connection,
- Positive locking connection: serration, polygon profile connection, profile shaft, connection, feather key connection, snap-on connection, riveted joint connection, shrinkage connection, pin connection, crimp connection,
- Metallic continuity: welding, soldering, gluing, vulcanizing

Avoidance of a mechanical contact,
- Absolute: ensure safety clearances are maintained between two parts (heat radiation),
- Relative: ensure an operating envelope between moving parts and static parts

Sealing area:
- Static sealing: O-ring, gasket, liquid sealant,
- Dynamic sealing: radial shaft seal, axial shaft seal.

Tool access:
- Sufficient space for a mounting,
- Sufficient space for a dismounting tool

Transport safety:
- Blind plug of tube,
- Seal lock for bores and electrical connections,
- Protection caps for shafts.

Under column physical contact it might be beneficial to mention also necessary information so that a functional performance will not be disturbed. A typical example could be a minimum safety distance between two parts because of heat convection or a potential overlap under dynamic movement.

Energy Transfer

Every interface with a transfer of energy between two parts can be described with well known forms like:

Mechanical energy:
- Kinetic energy,
- Potential energy and it's derivatives,
- Oscillation energy,
- Elastic energy,
- Sound energy.

Thermal and internal energy

Electrical and magnetic energy:
- Electrical energy,
- Magnetic energy,
- Electromagnetic energy.

Bond energy:
- Chemical energy,
- Nuclear energy (not in the automotive industry).

Often there is an energy transformation from one state of energy into another. Examples may be a transfer from electrical energy into mechanical energy or from chemical energy into thermal and mechanical energy.

In addition to that interface category there may be a condition where an active energy transfer needs to be avoided, so that a function of a system is not disturbed (e.g. a heat shield between two parts).

Information exchange

This category summarizes all human-machine-communication items as well as signal and data transfers between two controllers. Some typical examples are shown in the following list.

Check and control of preventive maintenance units:
- Coolant fluid level,
- Brake fluid level,
- Windscreen cleaning fluid,
- Headlamp cleaning fluid,
- Oil level check,
- Tire pressure check.

Identification or operation of elements:
- Position of reverse gear on shift knob,
- Opening / closing of fuel filler flap or engine hood,
- Adjustment of seat geometry,
- Operation of car jack.

Signal input and output of controllers / sensors:
- CAN-Bus-Signal,
- Analog signal of sensor,
- Passive driver safety (e.g. ABS, ESP, Airbag).

Driver warning:
- Optical warning light,
- Error message with display,
- Acoustical warning signal,
- Ice warning,
- Hand brake activated,
- Collision warning (parking aid).

Driver information:
- Mileage with fuel,
- Fuel consumption,
- Speed indication,
- Motor rotation speed,
- Cooling fluid temperature,
- Engine oil pressure,
- Internal / external temperature,
- Navigation.

Driver communication:
- Mobile phone,
- Blue tooth,
- Safety message,
- Internet access.

Passenger entertainment:
- DVD,
- Games.

Failure diagnostic,
- Readout of failure codes at dealer,
- Back up failure code,
- Programming of new software.

Material Exchange

All connections that are responsible for a material transport of media, liquids or materials are part of this category. Typical automobile applications are:

Powertrain Engineering:
- Oil cooling of engine and transmission,
- Transport and guidance of exhaust gas,
- Engine cooling fluid,
- Air duct for suction system,
- Additive diesel particulate filter.

Chassis Engineering:
- Fuel flow (Diesel, Petrol, Bio-Diesel, Natural Gas, Ethyl Alcohol, Bi-Fuel),
- Steering gear fluid,
- Brake fluid,
- Tire medium (Air, Nitrogen),
- Shock absorber oil,
- Grease lubrication.

Body Engineering:
- Windscreen cleaning fluid,
- Refrigerating agent of climate control.

INTERFACE MANAGEMENT

When integrating mechanical and electrical components in a vehicle there is a necessity to control that all respective interfaces will interact and cooperate simultaneously. To demonstrate this with the relevant quality tools like a boundary diagram, an interface matrix and an interface description sheet, a personal digital assistant (PDA) with integrated vehicle navigation control has been selected as a case study for a detailed investigation [3].

In the past built-in navigation systems for car application have been applied mainly with a high cost, but now there are portable systems available which are ready for use within minutes of installation.

Boundary Diagram of a Personal digital Assistant (PDA)

In a first step a boundary diagram is established to show all functional linkages of the PDA in a vehicle (Fig. 2). This kind of diagram describes in an abstract illustration the mayor components of the system and their combination with all neighboring parts. The importance is to distinguish a clear system boundary between inner and outer elements.

The PDA will be fixed with a flexible removable mechanical linkage in the car so that there is a good visibility for the driver. The power supply will be arranged by a lead connected to the cigarette lighter. As a base prerequisite the GPS-Antenna (Global Positioning

System) needs four signals from GPS-Satellites to determine the actual location. The driver can use a stylus and the tough screen to adjust the instrument and to give in the necessary destination information. The actual street information will be presented visually over the display and or/over a loudspeaker. With an external computer it is possible to install additional software and digital roadmaps to the memory card of the PDA. The fundamental interaction of the main components is demonstrated in the simplified boundary diagram. As shown in the diagram there are 17 single components which result in nine direct interfaces.

Interface Matrix for a PDA

The interface matrix is a recommended robustness tool of FORD [1] and is a prerequisite before starting a design FMEA. The principal construction is very similar to a distance table between main cities, which can be found in almost every road map or atlas (Fig. 3). With an interface matrix all system interactions must be identified including their strengths. This structured procedure requires the analysis of all inner to outer systems or components within the four quadrant categories. This is of a high importance as a non-recognition of strength could lead to potential warranty claims or vehicle recalls. With this background one recommendation is to create an interface matrix especially for newly designed constructions or when implementing new technology. With this matrix it can be detected if there is a productive or counterproductive relationship between two components. The example PDA shows a result of 136 potential combination possibilities with the 17 elements.

When considering an Interface the presented four quadrant approach expresses the strength of an interface value by five discrete figures (+2; +1; 0;-1;-2). Behind these figures there is following definition:

+2	Interaction is necessary for function
+1	Interaction is beneficial, but not absolutely necessary for function
0	Interaction does not affect function
-1	Interaction causes negative effects but does not prevent function
-2	Interaction must be prevented to achieve function

If the analysis shows four times a zero in this four quadrant description, the logical consequence is that the effected parts do not have any relationship with each other. A combination of one element with its own also does not give any additional knowledge so that the diagonal in the matrix will be left blank.

In a first step all internal combinations which are part of in the inner system boundary (yellow colour) will be analyzed in detail (Fig. 3). After the identification of all inner interfaces the next step is to concentrate on the direct interfaces (blue colour). As shown in the matrix there are forty potential combinations as compared to the obvious nine links when considering the boundary diagram only.

The third step (green colour) reflects the combination of all external elements with each other. After a fully investigation of the combination possibilities the next step is to take a detailed look at the interfaces with an interface description sheet.

Interface Description Sheet for a PDA

This document will describe the type of interface (Fig. 4) and the detailed linkage related whether it is a physical contact, an energy transfer, a material exchange or an information exchange. Therefore all functional targets including their tolerances will be listed. The column "technical specification" can be used to list all effected engineering specifications. The importance column is used to express which impact will result on the performance of a function. This might be beneficial for clarification of the necessary priorities. The column "impact" can describe if an interface input or output can result in a positive (desired) or negative (undesired) effect.

A further advantage of this description sheet is the declaration of interface partners by name as every component part has a department which is responsible for the part's interactive function. If a system needs to be integrated with different departments this document is very useful at technical meetings, along with technical drawings. This sheet can also be used as a tracking list which can be updated on a regular base to gather and agree on technical details with all effected parties during the development phase. Open issues can quickly be detected and discussed.

COMBINATION WITH ADDITIONAL QUALITY TOOLS

Boundary diagrams, interface matrices and interface description sheets are preventive quality tools which are ideally carried out before starting a design-FMEA as according to the QS-9000 standard. The target of all these documents is to distinguish the analyzed system and to structure all related interfaces. The gained knowledge can be incorporated into a design-FMEA. In order to calculate the impact of noise factors it is possible to additionally create parameter diagrams (P-Diagrams) [2].

CONCLUSION

The precise and exact control, calibration and adjustment of numerous vehicle interfaces has become mandatory and has increased the demand on technical departments today and for the future. Attached is a combination of the presented quality tools which are very well adapted for demonstrating an effective system integration. Besides the parameter diagram [2] these are further quality tools which demonstrate gained engineering knowledge. These documents should be used in technical meetings on a regular base. Relevant detail information will not be ignored or missed as this methodology demands the consideration of all interface combination possibilities and their peripheries. This leads to a significant contribution for failure-free and fail-safe products.

ACKNOWLEDGMENTS

The author wants to say to thank you very much to Joe Bakaj, Michael Droste and Hans Droemer for the official Ford Management approval of this article.

Many thanks also Peter Maher for the detailed reviewing and his comments to this article.

REFERENCES

[1]   FMEA Handbook Version 4.1, February 2004, (FORD internal)

[2]   Fritzsche, R.: Using Parameter-Diagrams in Automotive Engineering Application-Criteria, Guidelines and Best Practise; ATZ Worldwide; Volume 108, June 2006, p. 17

[3]   Mobiles Pocket PC Navigationssystem MD 95000, Instruction Manual, Medion

[4]   Pahl, G.; Beitz, W.: Konstruktionslehre, Methoden und Anwendung, 4. Auflage, Springer Verlag, 1997

[5]   Steinhilper, W.; Sauer, B.: Konstruktionselemente des Maschinenbaus, Springer, Berlin; Auflage: 6., Korr. Nachdruck, Januar 2005

CONTACT

Dr.-Ing. Ralf Fritzsche

Ford Motor Company

John-Andrews-Development Center

Spessartstrasse

D-50725 Koeln / Germany

Phone: +49(0)221/ 90-35056

Email: RFRITZSC@FORD.COM

DEFINITIONS, ACRONYMS, ABBREVIATIONS

ABS	Anti-lock Braking System
CAN	Controller Area Network
DVD	Digital Video Disc
ESP	Electronic Stability Program
FMEA	Failure Mode and Effects Analysis
GPS	Global Positioning System
PDA	Personal Digital Assistant

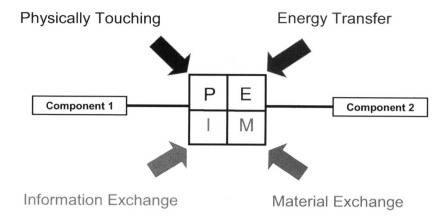

Fig.1 : Four quadrant structure of an Interface

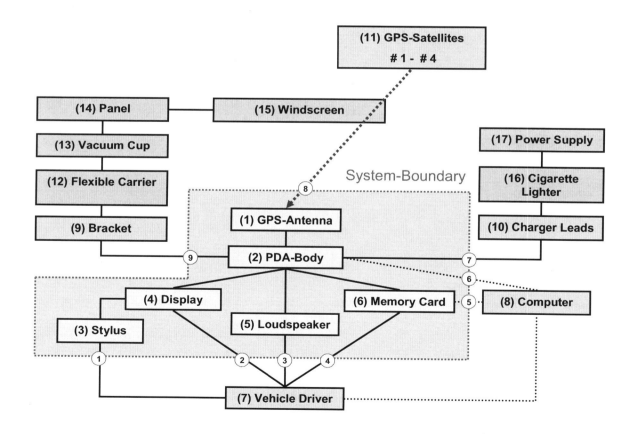

Fig.2 : PDA-Example of a Boundary Diagram

Fig.3   : PDA-Example of an Interface Matrix

Navigation with PDA in vehicle

Row/column labels:
(1) GPS-Antenna
(2) PDA-Body
(3) Stylus
(4) Display
(5) Loudspeaker
(6) Memory Card
(7) Vehicle Driver
(8) Computer
(9) Bracket
(10) Charger Leads
(11) GPS-Satellites #1 - #4
(12) Flexible Carrier
(13) Vacuum Cup
(14) Panel
(15) Windscreen
(16) Cigarette Lighter
(17) Power Supply

Potential combination possibilities: 0, 1, 3, 6, 10, 15, 21, 28, 36, 45, 55, 66, 78, 91, 105, 120, 136

Sum of potential combination possibilities: 15, 40, 81, 136

Sum of relevant combination possibilities: 7, 20, 20, 47

From	To	Interface / Flow	Category	Target Range	Unit	Description	Technical Specification	Importance	Impact	Interface Partner	Status	Comment
Vehicle Driver -> PDA												
(7)	(4)	Brightness of display	E, I	150-200	[cd/m²]	Adjustment of brightness for night and day	DIN EN ISO 9241	high	+	Supplier of tough screen	Green	none
(7)	(3)	Hand operation with stylus of display	P	5 +/- 0,1	[mm]	Diameter of stylus		medium	+		Green	none
(7)	(3)	Hand operation with stylus of display	P	110 +/- 0,2	[mm]	Lenght of stylus		medium			Green	none
(7)	(5)	Loudness of loudspeaker	E, I	30 - 60	[dB]	Adjustment of loudness		medium			Green	none
(7)	(2)	Switch-On/switch-off of PDA	E, I	2 +/- 0,1	[N]	Contact pressure		medium			Green	none
PDA -> Vehicle driver												
(5)	(7)	Acoustical information	I	40 - 70	[dB]	Replay	ISO 226:2003	high	+		Green	none
(4)	(7)	Optical information	I	300 - 600	[lux]	Onscreen dissolving		high	+	Supplier of touch screen	Green	none
(4)	(7)	Optical information	I	4000 +/- 10	[mm²]	Tough screen surface		high	+	Supplier of touch screen	Green	none
Charger Leads -> PDA												
(10)	(2)	Electrical power transmission	E	5 +/- 0,2	[V]	On-Board power supply voltage of PDA		high	+	Vehicle manufacturer	Red	Contact meeting with OEM's
(10)	(2)	Electrical power transmission	E	1 +/- 0,05	[A]	On-Board power supply voltage of PDA		high	+	Vehicle manufacturer	Red	Contact meeting with OEM's
Bracket -> PDA-Body												
(9)	(2)	Mechanical fixation and positioning of PDA	P	70 +/- 0,5	[mm]	Lenght of PDA take-up in x-direction		medium	+	Supplier of plastic material for bracket	Yellow	Meeting on 24.12.2006
(9)	(2)	Mechanical fixation and positioning of PDA	P	16 +/- 0,5	[mm]	Width of PDA take-up in y-direction		low		Supplier of plastic material for bracket	Yellow	Meeting on 24.12.2006
(9)	(2)	Mechanical fixation and positioning of PDA	P	25 +/- 0,5	[mm]	Height of PDA take-up in z-direction		medium	+	Supplier of plastic material for bracket	Yellow	Meeting on 24.12.2006
PDA-Body -> Bracket												
(2)	(9)	Mechanical fixation of PDA	P	70 +/- 0,2	[mm]	Lenght of PDA body in x-direction		medium	+	Supplier of PDA-body	Yellow	Meeting on 24.12.2006
(2)	(9)	Mechanical fixation of PDA	P	16 +/- 0,2	[mm]	Width of PDA body in y-direction		medium	+	Supplier of PDA-body	Yellow	Meeting on 24.12.2006
(2)	(9)	Mechanical fixation of PDA	P	25 +/- 0,2	[mm]	Height of PDA body in z-direction		medium	+	Supplier of PDA-body	Yellow	Meeting on 24.12.2006
GPS-Antenna -> GPS-Satellites												
(1)	(11)	Reception of GPS-signal	I	1575,42	[MHz]	L1-Frequency: Coarse/Acquisition-Code of GPS		high	-	US ministry of defence	Green	Civil usage of signals guranteed
(1)	(11)	Reception of GPS-signal	I	1227,60	[MHz]	L2-Frequency: P/Y-Code to compensate ionospheric effekts of GPS		high	-	US ministry of defence	Green	Civil usage of signals guranteed
Computer -> PDA-Body												
(8)	(2)	Datatransfer with external computer	I	min 56	[kbit/s]	USB Slave - wire fixed synchronisation with computer		high			Yellow	Diskussion about data interface
(8)	(2)	Datatransfer with external computer	I	min 56	[kbit/s]	Data loading on memory card		high		Supplier of memory card	Green	Use of SD or MMC-card format
Vehicle Driver -> Windscreen												
(7)	(15)	Visibility through windscreen	I	visible surface	[mm²]	Visible surface through windscreen shall not be distrubed by any parts of PDA		high		OEM	Green	Comment in instructionmanual about correct positioning at windscreen
Panel -> Windscreen												
(14)	(15)	Temporary fixation of PDA in vehicle interria	P	0,8 - 0,9	[bar]	Contact pressure on windscreen (Adhaftdruck am der Windschutzscheibe (Adhesion force)		high		OEM	Green	none

Fig. 4 : PDA-Example of an Interface Description Sheet

Advancements in Hardware-in-the-Loop Technology in Support of Complex Integration Testing of Embedded System Software	2011-01-0443 Published 04/12/2011

Andreas Himmler and Peter Waeltermann
dSPACE GmbH

Mina Khoee-Fard
General Motors LLC

ABSTRACT

Automotive technology is rapidly changing with electrification of vehicles, driver assistance systems, advanced safety systems etc. This advancement in technology is making the task of validation and verification of embedded software complex and challenging. In addition to the component testing, integration testing imposes even tougher requirements for software testing. To meet these challenges dSPACE is continuously evolving the Hardware-In-the-Loop (HIL) technology to provide a systematic way to manage this task.

The paper presents developments in the HIL hardware technology with latest quad-core processors, FPGA based I/O technology and communication bus systems such as Flexray. Also presented are developments of the software components such as advanced user interfaces, GPS information integration, real-time testing and simulation models.

This paper provides a real-world example of implication of integration testing on HIL environment for Chassis Controls. This application involves the integration of several ECUs such as powertrain, chassis and body electronics. It highlights the integration of plant models required for closed loop, running at different rates. The impact of serial communication on integration testing will also be discussed.

The paper concludes with how the HIL technology development is a continuous task that evolves in lock-step mode with the technical advancements of the automotive industry.

INTRODUCTION

Virtually all innovations in the field of automotive engineering are now based on new or further developed electronics. This applies to the latest engine technologies such as start-stop and hybrid drives, as well as passive and active safety systems (airbags, ESP, pre-crash), driver assistance systems (ACC, lane departure, park assist, night vision), and infotainment systems (telephone, internet, DVD player, TV).

Today's luxury vehicles include up to 70 ECUs that are connected with one another via several communication buses. The complexity of this networked structure of functions, combined with an enormous number of different vehicle variants (different engines and transmissions, different options, country-specific versions, etc.), is a major challenge in terms of developing and testing vehicle ECUs.

Hardware-in-the-loop technology (HIL) has become an integral part of the electronics development process of vehicle manufacturers and automotive suppliers for testing single ECUs as well as networked ECUs [7]. Single ECUs are frequently tested thoroughly by their suppliers before they are delivered to the car maker - the original equipment manufacturer (OEM) - for integration testing. The OEM then performs comprehensive subsystem and system tests with networked ECUs.

OEMs have the overall responsibility to ensure the safety and quality of a vehicle. They have a vital interest in the best quality of their products. Therefore, HIL integration testing of vehicles has become the proven technology to ensure the quality of complex networks of electronic control units. Integration testing is required to test the correct interaction of all the ECUs from different suppliers before production start of a new vehicle.

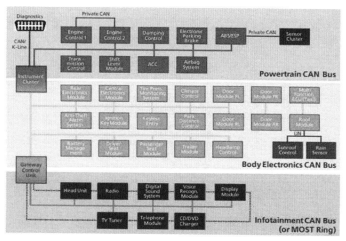

Figure 1. Typical electronics communication architecture of a modern vehicle [6]

Integration testing includes test of a certain function within a domain (e.g. chassis controls); integration testing of certain features which cross multiple domains (e.g. heated seats); and system level tests for the entire vehicle. Integration testing is based on subsystem and system level requirements in contrast to component level testing which has component level requirement sets.

This paper gives an overview about Hardware-in-the-Loop integration testing and its advancements. First the seamless and efficient use of multi-core processors for multi-processor systems is described. Subsequently the application of FPGA based I/O boards is described including an example on test of hybrid systems. The latter especially benefit from FPGA based I/O boards. Concerning software for integration tests a special emphasis is put on experiment software and real-time testing. A short overview on models is also given for completion. The comparison of component level vs. integration testing for chassis controls is shown based on an example on brake controls testing.

CHALLENGES OF INTEGRATION TESTING

HIL integration testing is the only way to test the proper functionality of hundreds of vehicle features with multiple ECU network protocols for a vehicle. Integration testing has to cover networked ECUs with variants existing due to platform strategy, country-specific versions of a vehicle or selectable vehicle options by customers. Likewise, integration testing needs to cover the complexity of vehicle networks consisting of up to 70 different ECUs. Thus, HIL integration testing requires covering large number of ECUs networked with a number of different buses and protocols covering large numbers of I/O signals. Networks for current vehicles most often use CAN, LIN, FlexRay, or MOST while Ethernet will be seen more and more in the future.

A majority of functionalities and features in modern vehicles depend on a number of networked ECUs. To test networked ECUs is typical for integration testing. For instance, integration testing for electric park brake (EPB) involves 6 ECUs within the vehicle. The ECUs affecting the integration testing are EPB control unit, electronic brake controller, engine controller, and three other ECUs within body and display area. In order to test ECUs for a hybrid powertrain the overall vehicle control unit (VCU) supervising the cooperation of engine and ECUs for electric drive(s), transmission, battery and braking system needs to be considered.

Testing of networked functionalities requires testing sub-networks of the total vehicle network, i.e. sub-networks representing complete vehicle domains. This is effective, since most networked functionalities are located within one vehicle domain. Depending on the test strategy of an OEM, the complete vehicle network might also be required. The loop time required for integration testing of vehicle domains differ. Short loop times (max. 1ms) are usually required for the powertrain and safety critical domains:

• Powertrain: engine, transmission, hybrid drives

• Chassis: ABS, ESP, driver assistance systems

• Safety: airbags, adaptive cruise control (ACC)

Testing of body electronics requires relaxed loop times (1 ms is for most cases more than sufficient) with synchronous execution of models and the I/O lines, due to diagnostic functions. Features and functions within body domain for integration testing demand a closed loop for accurate functionality of features with end to end latency requirements. For example, movement of a power seat by actuating a switch and positioning of the seat to desired location requires a closed-loop setup. In contrast, component level testing has more relaxed loop times in body domain.

Complexity of networked ECUs using various communication protocols poses a challenge for both component level and integration level testing when one or more ECUs are not present during testing. Therefore, rest bus simulation strategy (RBS) is used to simulate the serial communication signals from one or several ECUs in the system. Specifically, in component level testing where the focus is verification of operation of one ECU, the surrounding environment is simulated through RBS. Integration testing of networked ECUs within a domain also requires RBS. At initial stages of software development, one or more ECUs might not be available to the user. At the later stages of development when all ECUs within a subsystem are available, RBS might still be needed when cross domain interaction is required. For example, within body domain, the engine control module is not present and many functions within body domain require "engine speed" signal. This can be easily simulated using the rest bus simulation.

HARDWARE FOR HIL SYSTEMS

HIL systems for variety of different applications of the automotive industry must meet a number of essential requirements. Basically, they must be reliable, adaptable to every application and scalable from component to subsystem to system level.

Figure 2. Scalability of HIL simulator systems: From HIL I/O boards to desktop, rack and multi-rack systems

Thus, they must be based on a modular hardware system that offers the ability to adapt the processing power, the number of I/O channels for signal generation and measurement including bus interconnects, and offer all I/O and bus interfaces used in the automotive industry. It is necessary to build various sizes of simulators with a modular system ranging from table-top single boxes with one HIL I/O board up to large systems consisting of several networked simulators.

Beyond the technical advances of a modular system it also enables investment protection, since it enables a modification of an HIL system to adapt it to new project requirements. The HIL systems consist mainly of off-the-shelf (OTS) components/products with long-term availability and long-term software support. These OTS components are integrated to turnkey HIL systems for integration testing.

MULTI-CORE PROCESSOR BOARDS AND MULTI-PROCESSOR SYSTEMS

In hardware-in-the-loop systems, multiprocessor (MP) systems have long been commonly used to scale both processing power and I/O performance. Of course, MP systems for long time profited from the advances of the processing power of CPUs. The usual method of boosting processor speed in the past was to increase the clock rate or to improve processor architecture. While increasing clock speeds are coming up against its physical limits, there remains very little room for further improvement for the latter method. Multi-core (MC) processors, i.e., ones with several CPU cores, are the way out of this dilemma. Each of the CPU cores has greater performance than earlier single-core processors, and the rapid data communication between the cores is a real bonus.

The challenge of applying the processing power of multi-core processors for real-time simulation therefore lies in thinking out how to handle all the different tasks, in other words, how to most usefully distribute and parallelize them, and how to organize communication between them.

Definitely, a method to interconnect multi-core processors needs to fit seamlessly into the concept of MP systems for HIL systems. The obvious advantage of MP systems is the independent access of each processor to its connected I/O via short, low-latency, broadband connections. The clear separation of processors in typical MP systems is particularly useful. Moreover, MP systems allow modularity. For example, test systems for single ECUs or vehicle domains can be interconnected to produce network simulators.

The procedures commonly used in multi-processor (MP) systems can be applied to the individual cores in multi-core processors. This approach is described. It is shown, that for inter-core and inter-processor communication the same user interface and communication principles can be used, although the underlying technical implementation can be quite different. This results in a consistent buildup of systems from small multi-core systems to large multi-processor systems. Apart from the advances for the HIL user and technical benefits of this approach, it also enables investment protection due to the fact that upgrade of existing HIL systems with a new powerful processor board is a simple task. A special benefit of the MC/MP approach is the modularity it enables for the real-time models.

Seamless Inter-Core and Inter-Processor Communication

In order to achieve this modularity seamless inter-core and inter-processor communication links are required, as mentioned above. Figure 3 shows an example of a small MP/MC system visualizing the seamless communication principle of the MC/MP approach. The system consists of two processor boards with a quad-core processor on each of them. Each board has 4 ports for inter-core communication (GL0 ... GL3). Likewise, each core has 4 ports (GL0 ... GL3). These can either be connected to ports for inter-processor communication or connected to inter-core communication links (dashed lines). The latter have the same functionality than inter-processor links. The topology of this network is totally user configurable.

Inter-processor links shown in Figure 3 are based on optical data transmission via optical fiber connecting ports on distinct processor boards. This is a proven technology applied for a very large number of HIL simulators built upon single-core processor boards and in use throughout the world. In contrast to this inter-core communication links are implemented in software utilizing data exchange through the cache memory of the processor. As a result inter-core links have a very high bandwidth compared to optical inter-processor links. Furthermore, their latency is significantly lower.

This concept of seamless inter-core and inter-processor communication enables multiprocessing using a number of cores of only one CPU as well MP systems consisting of several processor boards. In addition, a mixture of multi-core and single-core processor boards is possible. For an efficient use of this concept the software support and user interface is important.

Data exchange between tasks on single-core processors in MP systems can be configured graphically in Simulink® by means of inter-processor communication (IPC) blocks (multiprocessor extension of the Real-Time Interface (RTI-MP) software) for a long time. A new version of this software can now also be used to configure inter-core and inter-processor communication in the same environment, although the underlying communication has physically different

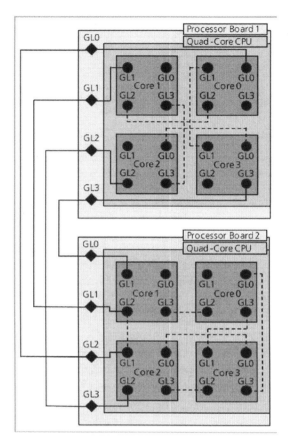

Figure 3. Configurable interconnects for inter-core (dashed lines) and inter-processor (solid lines) communication in an MP system consisting of two processor boards with quad-core processors

implementations. This ensures a simple scaling of computing power to user-defined systems of nearly any size since RTI-MP encapsulates the physical layer of communication. An optimized use of the given computing power is enabled by a user defined network topology depending on the use-case. Closely associated with this is an easy migration of single-core multi-processor HIL systems to multi-core HIL systems. With this it is easy to benefit from very high bandwidth of inter-core communication. The reasonable maximum number of I/O connected to one CPU certainly needs to be taken into account.

When using HIL simulators based upon a seamless MC/MP system interfaces for real-time models can be identical with inter-core and inter-processor communication links due to the abstraction of the physical layer by the configuration software. This enables an easier exchange of models, such as managing different variants of engine models, handling of country-specific variants, handling of low to high-end configurations of the vehicle, all with the same HIL environment. In addition, model exchange with suppliers is made easier.

The following example is to show the application of an HIL system with inter-core and inter-processor communication links for integration testing. Figure 4 shows the overall Simulink® real-time model for integration testing of an electric vehicle. The overall model consists of application dependent plant-models for the domains of the vehicle and the bus communication: vehicle dynamics, electric drive, battery management, power steering, chassis control, and simulation of serial communication. This structure of the model is identical to the MC/MP structure of the HIL simulator, as shown in Figure 5. Thus there is a direct mapping. In addition to the plant models the overall Simulink® real-time model also comprises the MP/MP configuration. When compiling this overall model executables are generated and automatically downloaded to the assigned cores of the MC/MP system.

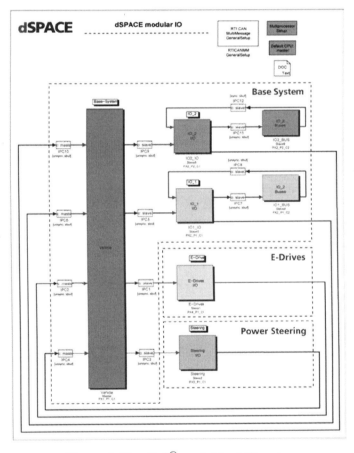

Figure 4. Simulink® model for MP system

To adapt an MC/MP HIL system to different use-cases model parts can be exchanged. The ECU for a 4 cylinder engine and the ECU for a 6 cylinder engine need to be tested with the same HIL system. A transition between these use-cases only requires an exchange of the appropriate real-time model, which is assigned to only one core of the HIL simulator. It is not necessary to change the hardware of the simulator, since every core of a real-time processor can have its own I/O.

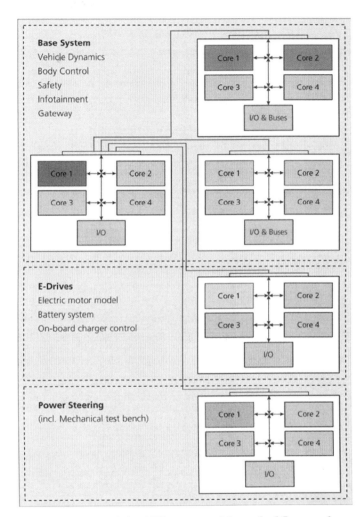

Figure 5. Modular HIL system with total of five quad-core processor boards, the functions of the different cores and the associated I/O hardware

I/O AND BUS BOARDS - SIGNAL GENERATION AND MEASUREMENT

For HIL systems, processing power is essential, but high I/O throughput in combination with integrated signal conditioning is also key. The I/O and bus boards of a modular system have to cover the signal generation and signal measurement of a broad range of applications like, engine, transmission, electric vehicle, vehicle dynamics, and vehicle interior systems. These differ at least partly depending on the vehicle type: passenger cars, light and heavy-duty trucks, racing cars.

Typical standard HIL I/O boards need to be equipped with multi I/O along with signal condition to handle the typical I/O channels of automotive ECUs (e.g. digital I/O, analog, resistive simulation, PWM I/O …). They also need to have a CAN bus interface. To test ECUs for combustion engines standard HIL I/O boards need to handle usual engine-angle-related signals (e.g. crank, cam, ignition, injection). Some use-cases require fast I/O subsystems for special applications, such as engine knock and active wheel speed signal generation by means of digital signal processors (DSP). I/O boards meeting these demands have been available for a long time [6] as well as I/O boards having a number of I/O channels of the same kind. Boards with a larger number of digital I/O channels and signal conditioning are required to test subsystems such as chassis systems. Other well known examples are I/O boards to handle communication buses (e.g. for CAN, LIN, FlexRay) to test bus communication of single or networked control units.

Some fields of application (e.g. electric machines) require demanding processing of I/O signals within short cycle times. These requirements are met by application of FPGA technology for I/O boards. The basic principle is to transfer parts of the plant model from the CPU to an FPGA on the I/O board. This enables very short processing times locally on the I/O board that are unaffected by transmission times via an I/O bus to a real-time processor. User friendliness is crucial to utilize these technical benefits efficiently. These I/O boards can be adapted to different use-cases due to flexibility provided by FPGA technology.

In order to ensure user comfort a HIL supplier provides out-of-the-box solutions for FPGA I/O boards. The HIL user can make use of the benefits of FPGA technology without having to worry about programming it. This is a special advantage, since FPGA programming requires different skills, knowledge and tools than programming real-time simulation models. Examples of out-of-the-box solutions are described in the subsequent section on HIL tests of hybrid drives.

In contrast to out-of-the-box solutions it is also possible to give the user of HIL hardware full control of FPGA I/O boards. Figure 6 shows the workflow to develop FPGA applications to be used in combination with processor-based real-time applications. The development steps required for the FPGA are shown on top ("FPGA development path"). Below are the usual steps to develop a realtime application ("Function developer path"). Both paths start with the development of the appropriate part of the plant model. Simulink® is usually used to develop a real-time model. Special tools exist to develop FPGA applications, like the Xilinx Design Suite. Second step is to build each model, i.e. to compile the Simulink® model and to synthesize the FPGA

application. Up to this point both parts of the workflow do not need to be coupled. Thus different specialists can work independently of each other using different development tools. At the end both applications are loaded to the Hardware-in-the-Loop system and executed. Optionally it is possible to test both applications in combination without using a Hardware-in-the-Loop system. This is called "off-line simulation". Once a FPGA application is developed, synthesized, and successfully tested it can be stored in a library to reuse it.

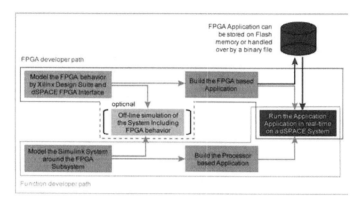

Figure 6. Workflow for the development of an FPGA application

To interface the user-developed FPGA application from the real-time application dedicated FPGA code is required. This is developed and supplied by the HIL vendor in combination with the FPGA I/O board. He also supplies specific FPGA code required to access the I/O ports implemented on the FPGA I/O board from the user-developed FPGA application.

EXAMPLE: HIL TESTS OF HYBRID DRIVES

A special application field is HIL tests of control units for the drivetrain of hybrid and electric vehicles. Tests on electric motors require the simulation of high currents, fast reactivity and simulation of sensor signals of electric motors.

Figure 7 shows the interfaces that are usually applied to connect an electric machine to an HIL system. However, the preferred test interface for ECUs of electric drives in hybrid or electric vehicles is the signal level interface between the controller and the power output stage. The usage of the signal level interface offers the best flexibility to adapt the HIL simulator to new devices under test. In addition the necessity to handle high currents or even a complete electric machine is eliminated. Having high current or complete electric machine in the lab environment requires special operational safety requirements to be met.

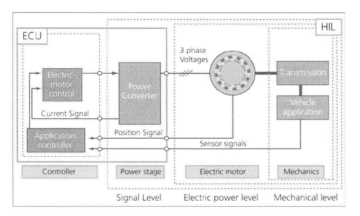

Figure 7. Interfaces for electric motor simulation

To test electronic control units for full-hybrid passenger cars the simulation of at least two electric drives (each of these with a 3-phase voltage) by an HIL system is of special importance. The reason is the market relevance of power-split hybrid drives. Two electric motors are also used for an advanced version of the power-split hybrid drive, the two-mode hybrid drive. The latter is completed by four gears with fixed gear ratio to optimize the efficiency and power output for a broad range of drive situations.

A decisive performance factor for the simulation of electric drives is the minimum sampling time of the control loops achievable with an HIL system. This value is crucial for the stability and accuracy for the control of the electric drive.

This requires precise measurement and generation for the electric-motor control by the HIL simulator. On one hand all relevant timing information of the PWM signals (duration of the PWM signal, duration of the high- and low signals, and the dead times) need to be measured. On the other hand the signals for rotational speed and angular position need to be simulated. This requires the simulation of analog types of sensors (resolvers, encoders) and digital sensors (incremental encoders). In addition to these I/O requirements high-processing-power is needed to run the simulation models. A third prerequisite is a high-speed I/O bus having very low latency to connect the I/O with the processor board.

These requirements can be met with an HIL system based on the quad-core processor board and I/O solutions for electric drives based on FPGA-based I/O boards.

With such a configuration, one core of the processor simulates a model of the combustion engine and models for the two electric machines are running on two other cores. The forth core can then be used to simulate the bus communication of the control units for the combustion engine and the electric machines. It might also be used to simulate a third electric motor if required.

This system enables a time resolution of the measured PWM signals (center-synchronized measurement) better than 25 ns. Depending on the PWM frequency, the sampling times are between approx. 30-60 µs. The signals indicating rotational speed and angular position of analog and digital sensors can be simulated with a time resolution of 100 ns and 25 ns, respectively. These time resolutions are independent whether two or three electric motors are simulated.

SOFTWARE FOR HIL SIMULATORS

MATLAB®/Simulink® (ML/SL) has become established as a quasi-standard for function development in the automobile industry. Therefore it is logical to use ML/SL to describe both the dynamic behavior of the plant, and the definition and configuration of the I/O of a simulator system.

The entire function and instrumentation code for the real-time system is then generated by autocoding. Figure 4 shows the top hierarchy level of an HIL simulation model.

EXPERIMENT SOFTWARE

Today, numerous different special tools are used to develop modern automotive electronics. The tools provide access to simulation platforms for rapid control prototyping and hardware-in-the-loop simulation as well as to connected bus systems (such as CAN, LIN and FlexRay), and can perform measurement, calibration and diagnostics on ECUs via standardized ASAM interfaces.

The development tools for this are specially tailored to specific application contexts, and all too frequently, expert knowledge is needed in order to work productively. Now that application areas are beginning to overlap more, greater flexibility and integration are essential. There is an increasing need to access several different data sources, even though they might not necessarily have specialized knowledge of the overall system or of all the subdomains involved.

What makes mixed scenarios even more difficult is that different tools frequently have to be used in parallel. This derives the costs of software, training and operation higher, and makes interactive, automated operation more complicated. There is therefore a need for an experiment tool that optimally addresses the requirements of all the application scenarios and makes it easy to reuse data such as user interfaces, measurement data and parameter sets throughout the various development stages.

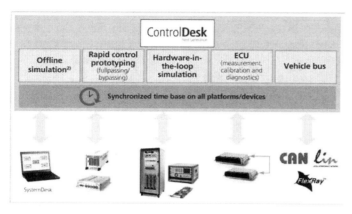

Figure 8. Universal experiment tool for ECU development

The required flexibility mentioned above demands a universal experiment tool for ECU development. It needs to provide a large number of interfaces for various application areas (ref Figure 8): Rapid-Control-Prototyping, Hardware-in-the-Loop simulation, ASAM MCD IMC (CCP und XCP), on-chip debug interfaces for ECU access, CAN- and K-Line-based diagnostic access via ASAM MCD 2D (ODX) with serial communication protocols such as KWP2000, UDS, TP2.0, and GMLAN. Thus there is a need to merge former experiment tools and calibration tools.

MODELS FOR HIL TESTING

To close the control loop between the ECU and simulator inputs/outputs, dynamic plant models are needed. These must provide a sufficiently good representation of the system to be controlled. The model fidelity must be sufficient such that the ECU does not detect any inconsistencies which would disable its operation. The real-time capability of the model is another decisive criterion for HIL use. Very detailed models (including FEM approaches and 3-D flow simulation) are often used in engine and chassis development, and such models are far from being real-time-capable. For HIL operation, therefore, specific models and model approaches are often used that are real-time-capable and with sufficient fidelity to satisfy the ECUs under test.

Combustion models: With engine simulators, mean value models are often used. These are usually generic and are parameterized by means of test bench measurements made for the specific engine. Static internal engine relationships are frequently approximated by using look-up tables. Either the engine models are operated in test bench mode (the engine speed is held constant by means of a simulated load machine) or drive cycles are run via a drivetrain model (e.g. EUDC, FTP75) [4].

Development work recently started on engines with in-cylinder pressure measurement, and this has made greater

model precision necessary for HIL operation. Therefore, models that describe inlet and outlet behavior and combustion very precisely are being used more and more. However, these models are far more complex to parameterize and also require smaller step sizes (typically 100 μs) [4]. Another current major development issue is the suitable simulation of the exhaust system with a particulate filter, an oxidation catalytic converter or even selective catalytic reduction (SCR).

Vehicle dynamics models: ESP ECUs are very safety-critical systems. Sensor signals are therefore monitored very precisely (e.g. by defined initialization sequences for each single sensor) and checked for plausibility toward one another by means of internal observer models. This results in tough demands on vehicle dynamics models. In HIL systems for vehicle dynamics control systems (ABS, ESP), the essential masses (car body and wheel mass) are simulated, and the wheel suspensions (interaction between the body and wheel) are represented by multidimensional look-up tables. The tire models, drivetrain and steering system are also extremely important.

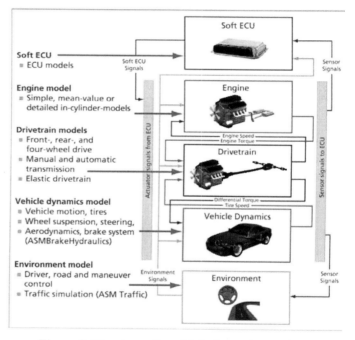

Figure 9. Structure of a vehicle dynamics model in MATLAB®/Simulink®

Environment models: As well as the actual vehicle models, virtual test drives also require environment models such as a driver model, a road description and a maneuver control (Figure 9). All these components have now become standard in commercially available vehicle dynamics models [5].

Currently, driver assistance systems (e.g. ACC, lane change/lane keeping/parking assistants, environment detection) are an important focus in the development of automotive electronics. Not only models of the main vehicle are required for these, but also models of the environment, in other words traffic simulation that appropriately simulates the behavior of other vehicles and the signals from radar sensors, ultrasonic sensors, and cameras, and feeds them to the ECUs.

Other development issues include different types of electric drives and their peripherals (vehicle electrical system, battery systems, DC-DC converters, etc.).

TEST AUTOMATION: REAL-TIME TESTING

The greatest benefit is derived from HIL systems if test runs are not performed interactively but automated. HIL tests frequently run overnight or on weekends, and all that the engineers can examine the generated test reports, identify failed tests, perform root cause analysis, and implement new tests. A 24/7 integration testing is standard at many of the OEMs worldwide.

Script languages such as VBA, MATLAB® and Python are frequently used for automation. However, there are also test tools on the market that allow test implementation to be performed in graphical form [3]. As in Simulink®, the test creator can put together a test from single test steps or whole test sequences from libraries, and "program" graphically parallel as well as serial test sequences.

Automated testing usually is performed by executing tests on a standard PC connected to the Hardware-in-the-Loop (HIL) system. However, this method is often not sufficient when greater timing precision is required - for example, if ECU interaction has to be captured and responded to in a range of milliseconds. The solution to this requirement is to develop the script in real-time Python. These scripts run on the processor board of the HIL system in real-time, i.e., synchronously with the model, so all test actions are performed on a real-time basis. Reactive tests which respond to changes in model variables within the same simulation step can be implemented this way. Time measurements in tests are also far more precise, as there are no latencies in communication. Simulation step size is now the only limit to the maximum time resolution of measurements.

A real-time Python interpreter, running on the processor board along with the model, allows the test script to be executed synchronously with the model. The interpreter can execute several real-time tests simultaneously and independently of one another. The tests can interact with the simulation model in real-time via the memory on the processor board. ECUs connected to the HIL simulator can therefore react within every individual simulation step.

It is common for real-time testing to use standard Python scripts from real-time testing libraries provided with the test automation tool (for example, for accessing model variables and executing several test branches within one real-time test in parallel). The user can also create its own Python libraries and reuse them in several tests.

Real-time testing is commonly used for vehicle integration testing within body domain subsystems. During the vehicle subsystem testing, a feature is tested for correct behavior within an integrated ECU environment. One of the examples is evaluation of power seats. Many of the test cases look for seat to be correctly positioned. The latency requirement from the time of actuation to the time of start of seat movement needs to be measured accurately. In this example, a typical power seat in a vehicle has seven degrees of freedom, three for position setting and four for lumbar setting. To position the seat accurately upon activation of switches, the realtime script is continuously reading the seat position until the desired settings are reached. In this case, "continuously reading" a variable is provided by real-time library which a user can use within test automation script.

COMPARISON OF COMPONENT LEVEL VS. INTEGRATION TESTING FOR CHASSIS CONTROLS

Integration testing poses additional requirements for HIL hardware and software. A HIL setup for component testing mainly focuses on software verification for a specific ECU under test. The software verification could be for performance or diagnostics check. The HIL hardware is generally configured to required I/O for testing ECU functions, diagnostics for a specific setup. Integration testing focuses on verification of a feature within a subsystem. Therefore, all ECUs affecting the functionality of a feature should be included in the HIL setup. The example described below, demonstrates the HIL setup differences for component level vs. integration level testing for chassis controls system with focus on brake controls testing. More complex examples describing the differences in component vs. integration testing can be found in [8].

A HIL setup for brake controls component level is shown in Figure 10.

The HIL setup for vehicle component and plant models consists of:

• Electronic brake control unit (EBCM)

• A sensor pack and a plant model to represent the brake hydraulics unit.

• Vehicle dynamics model

• Simulated inertial measurement unit (IMU) and steering angle sensor (SAS)

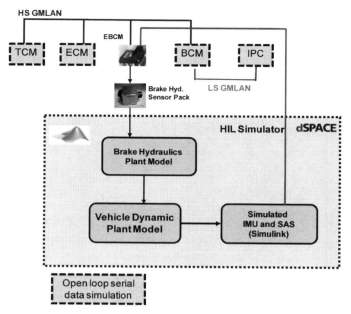

Figure 10. HIL setup for component tests of brake controls

A typical HIL system with this setup can only consume a small portion of throughput power of microprocessor, approximately 400 μsec.

In contrast, integration testing requires that all ECUs affecting the functionality of chassis controls to be present on the HIL bench setup. Individual subsystems may use subset of ECUs on the HIL bench. The example below focuses on brake controls integration testing for purposes of comparison with component level testing. In addition to verification of features and functions implemented in the controllers, integration testing verifies subsystem latency against the requirements, and serial communication and diagnostics. A typical bench setup for chassis controls integration testing is shown in Figure 11.

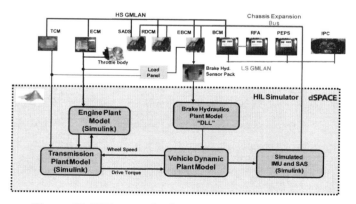

Figure 11. HIL setup for integration testing of chassis controls

The HIL setup requires up to 500 I/O signals for this subsystem. A multi-core processor is utilized to accommodate the required throughput for the subsystem testing. The ECUs used for integration testing of brake controls are:

• Engine Control Module (ECM),

• Transmission Control Module (TCM),

• Electronic Brake Controls Module (EBCM),

• Body Computer Module (BCM),

• Instrument Panel Control (IPC),

• Passive Entry Passive System (PEPS),

• Remote Function Actuation (RFA),

• Rear Drive Control Module (RDCM),

• Semi Active Damping System (SADS).

The plant models required for integration testing are described below. The integration of all required plant models requires enough throughput from HIL systems to ensure complete closed-loop environment within 1 msec loop frame.

• Engine and transmission plant models with appropriate level of fidelity for testing this subsystem.

• Vehicle dynamics model.

• Brake hydraulic model. This model is specific to suppliers of EBCM. In many occasions, these models are provided by suppliers in form of a Simulink® "S-function" or "DLL". It should also be noted that the brake hydraulic model is required to run at much higher rate than the rest of the plant models. In this example, the brake hydraulic model is running ten times faster the rest of plant models.

• The real-time implementation of serial communication has a significant impact on the throughput of the system [1]. In the case of chassis controls several communication protocols might be present (GMLAN high Speed, GMLAN Mid Speed, GMLAN Low Speed, CAN, and FlexRay).

• For the test automation of the brake controls subsystem, a camera for capturing the tell tales and messages on the IPC is also present in the setup. The required application setup for the camera is also integrated with the all plant models for this subsystem.

A benchmark study has been performed to study the throughput of the system when a single core processor is used vs. multi-core processor. First all models were compiled on one core and the throughput was measured. The system measured loop time is at 1.2 msec causing overruns in the system at the typical millisecond sample rate.

Later the HIL system was upgraded to a multi-core processor board and all models were compiled on one core of the multi-core processor. The system measured loop time improved by 30%. Table 1 outlines the impact of various models on the performance of the system on a single core processor. On average the percentage improvement for loop time is distributed evenly among various plant models.

These improvements on the throughput are a result of the multi-core processor board architecture efficiency in design.

Table 1. Loop times for plant models

Model Category	Approx. Time [μs]
I/O Model	310
Powertrain Models	50
Vehicle Dynamics	150
Brake Hydraulics (S-Function)	170
Serial Data Communication Blocks, Camera, Common Diagnostics Implementation	520
Total Required Throughput	1200

When using a multi-core processor in the HIL system setup, the models are divided into multi cores. The methodology used for dividing the model is not a trivial task. The approach generally starts with experimentation based on engineer's expertise and the analysis done on single core.

One of the strategies outlined below can be used for distribution of models on different cores:

• Use one core for "Serial Communication Blocks, Camera Setup, and Common Diagnostics". Remainder of models can reside on second core.

• Use one core for "Serial Communication Blocks, Camera Setup, and Common Diagnostics", second core for powertrain plant models, and I/O blocks, and the third core for vehicle dynamics model and brake hydraulics plant model.

The choice of strategy depends on many factors such as ability to have multiple chassis controls subsystems implemented on the same HIL bench. Therefore, the allocation of plant models into different cores needs to comprehend estimation of additional throughput requirements for future expandability. For example, the implementation of RDCM and SADS will incorporate additional S-functions which impact the throughput of the system. In addition the serial communication and common diagnostics blocks need to be expanded to incorporate these additional two subsystems.

SUMMARY/CONCLUSIONS

Hardware-in-the-Loop technology for integration testing needs to cover all aspects to verify features within a subsystem of ECUs, within a domain, or complete ECU networks of a vehicle. While well known essential technical requirements for HIL hardware are still valid (e.g. modular, and scalable system to build various sizes of simulators for a broad range of applications, off-the-shelf components, long-term availability and support), some new requirements arise for testing of new technologies being introduced to vehicles. Noteworthy requirements are the optimal use of processing power and signal generation and measurement within short cycle times. In order to use modern multi-core processors optimally for new HIL systems as well as to upgrade existing systems a concept of seamless and modular inter-core and inter-processor communication is presented. I/O boards for short cycle times (e.g. test of ECUs for hybrid drives) based on FPGA technology meet the requirements. In addition they can be used as out-of-the-box solutions or can be user-configurable. With respect to the software for HIL testing MATLAB®/Simulink® is the established defacto-standard for functional (behavior) modeling and development of dynamic plant models. To make optimal use of a HIL simulator within the development process the experiment software is evolving to a universal experiment tool for ECU development.

One of the demands on HIL software is to provide capability for real-time testing which can be used in conjunction with test automation tools. The real-time script libraries can be used with test automation tools or directly through real-time script languages (e. g. Python-RT).

REFERENCES

1. Lahdhiri, T. and Khoee-Fard, M., "GMLAN Real Time (GMLAN-RT) Simulation on dSPACE HIL Systems," presented at 2007 GM CAE Conference, USA, 2007.

2. Himmler, A., "Modular, Scalable Hardware-in-the-loop Systems," ATZelektronik worldwide 5(2):36-39, 2010.

3. Ziehr, L., Pascal, J., and Wiedemeier, M., "Control Validation Project at GM for Hybrid Vehicle Air Conditioning," SAE Technical Paper 2006-01-1446, 2006, doi:10.4271/2006-01-1446.

4. Schulze, T., Wiedemeier, M., and Schutte, H., "Crank Angle-Based Engine Modeling for Hardware-in-the-Loop Applications with In-Cylinder Pressure Sensors," SAE Technical Paper 2007-01-1303, 2007, doi: 10.4271/2007-01-1303.

5. Schutte, H. and Waltermann, P., "Hardware-in-the-Loop Testing of Vehicle Dynamics Controllers - A Technical Survey," SAE Technical Paper 2005-01-1660, 2005, 10.4271/2005-01-1660.

6. Wältermann, P., Schütte, H., Diekstall, K., "Hardware-in-the-Loop Testing of Distributed Electronic Systems", ATZ worldwide 05/2004, http://www.dspace.de/ftp/papers/dSPACE_ATZ_0405_e_f39.pdf.

7. Wältermann, P., "Hardware-in-the-Loop: The Technology for Testing Electronic Controls in Automotive Engineering" 6th Paderborn Workshop "Designing Mechatronic Systems", Paderborn, Germany, 2009, http://www.dspace.de/shared/data/pdf/paper/HIL_Overview_Waeltermann_03_E.pdf.

8. Khoee-Fard, M. and Lahdhiri, T., "Challenges in Real Time Control Simulation (Hardware-in-the-Loop) in Active Safety for Subsystem Level Software Verification," SAE Technical Paper 2011-01-0450, 2011, doi: 10.4271/2011-01-0450.

CONTACT INFORMATION

Dr. Andreas Himmler
Product Manager HIL Simulators
dSPACE GmbH
Rathenaustraße 26
33102 Paderborn
Germany
ahimmler@dspace.de

Dr. Peter Wältermann
Group Manager Engineering
dSPACE GmbH
Rathenaustraße 26
33102 Paderborn
Germany
pwaeltermann@dspace.de

Mina Khoee-Fard
Engineering Group Manager
General Motors, LLC
30003 VanDyke Avenue
Warren, Michigan 48090
mina.khoee-fard@gm.com

DEFINITIONS/ABBREVIATIONS

ACC
    Adaptive Cruise Control

ASAM
    Association for Standardization of Automation and Measuring Systems

BCM
    Body Computer Module

CCP
    CAN Calibration Protocol

CPU
    Central Processing Unit

EBCM
    Electronic Brake Control Module

ECM
    Electronic Control Module

ECU
    Electronic Control Unit

ESP
    Electronic Stability Program

FPGA
    Field Programmable Gate Array

GMLAN
    General Motor Local Area Network

HIL
    Hardware-In-the-Loop

I/O
    Input/Output

IPC
    Instrument Panel Control

KWP2000
    Key-Word-Protocol 2000

MC
    Multi-Core

MCD-2 D
    Data Model for ECU Diagnostics

MCD-2 MC
    ECU Measurement and Calibration Data Exchange Format

MP
Multiprocessor

ODX
Open Diagnostic Data Exchange

OEM
Original Equipment Manufacturer

PEPS
Passive Entry Passive System

PWM
Pulse Width Modulation

RBS
Rest Bus Simulation

RDCM
Rear Drive Control Module

RFA
Remote Function Actuation

SADS
Semi Active Damping System

TCM
Transmission Control Module

TP2.0
Vehicle Diagnostic Protocol

UDS
Unified Diagnostic Services

VBA
Visual Basic for Applications

VCU
Vehicle Control Unit

XCP
Universal Measurement and Calibration Protocol

About the Editor

John E. Blyler is vice president, chief content officer for Extensionmedia, which includes brands such as Chip Design, Solid State Technology, Embedded Intel, RF-Microwave Systems, and others. He is a frequent guest on Chipestimate. TV, where he covers technology trends and intellectual property issues. John was the senior editor for Penton's Wireless Systems Design magazine and, before that, an associate editor for the IEEE I&M magazine. He has co-authored several books on systems engineering and RF technology for Wiley and Elsevier. John has over 23 years of systems engineering hardware-software experience in the electronics industry—including work at the DoD, within the semiconductor private sector, and at software start-up companies. He remains an affiliate professor of graduate-level systems engineering at Portland State University. Mr. Blyler holds a BS in engineering physics from Oregon State University, as well as an MSEE from California State University, Northridge.

273TXT